TECHNOSCIENCE IN HISTORY

Transformations: Studies in the History of Science and Technology

Jed Z. Buchwald, general editor

A series list appears at the back of the book

TECHNOSCIENCE IN HISTORY

Prussia, 1750—1850

Ursula Klein

The MIT Press
Cambridge, Massachusetts
London, England

This book was set in Stone Serif by Westchester Publishing Services. Printed and bound in the United States of America.

Library of Congress Cataloging-in-Publication Data is available.

ISBN: 978-0-262-53929-6

10 9 8 7 6 5 4 3 2 1

CONTENTS

IV TOWARD NINETEENTH-CENTURY TECHNOLOGICAL SCIENCE

ILLUSTRATIONS

ACKNOWLEDGMENTS

This book project began almost fifteen years ago at my home institution, the Max Planck Institute for the History of Science in Berlin. It would have been impossible to carry out research without the help of my home institution and the support by Hans-Jörg Rheinberger, director of Department III, where the project began, and Jürgen Renn, director of Department I, where it was completed. I am grateful to Hans-Jörg Rheinberger and Jürgen Renn for their advice and their financial support of the project. Jürgen Renn's intellectual challenges and generosity have been an invaluable experience, which stimulated the extension of the project well into the period of the industrial revolution and the establishment of the modern technosciences.

Over the years I have also benefited from numerous discussions with my colleagues at the Max Planck Institute and other institutions, among them Bruno Belhoste, David Bloor, Matthew Eddy, Dieter Hoffmann, Peter Konečný, Marcus Popplow, Maria Rentetzi, Dagmar Schäfer, Matthias Schemmel, Florian Schmalz, Emma Spary, Matteo Valleriani, and Norton Wise. I am particularly grateful to Alan Chalmers, Jonathan Harwood, Wolfgang Lefèvre, Mary Jo Nye, Kathryn Olesko, Jürgen Renn, Alan Rocke, and Matthias Schemmel, who read the entire book manuscript or parts of it. This book has been vastly improved by their queries, suggestions, and generous help.

The project became feasible not least with the longstanding aid of my research assistant Johannes Lotze, who helped to track down archival sources and transcribed some 4,000 pages of archival material. I owe him an immense debt. I am also grateful to the archivists of the Archive of the Berlin-Brandenburg Academy of Sciences and Humanities, the Secret State Archives Prussian Cultural Heritage Foundation, the Archive of the Royal Prussian Porcelain Manufactory in Berlin, and the Autograph Collection of the Library of Humboldt University. Special thanks go to the librarians of my home institution, in particular Urte Brauckmann, Ellen Garske, Ruth Kessentini, and Matthias Schwerdt, for making rare books, articles, and illustrations available to me. Further, I am indebted to Anita Mage for her assistance with the translation of several parts of this book from German into English.

It is a pleasure to thank Jed Buchwald for including this book in his MIT book series Transformations: Studies in the History of Science and Technology, as well as Katie Helke, Susan Clark, and Laura Keeler from the MIT Press for organizing its publication. I have been particularly fortunate to have Matthew Abbate as my copy editor, and I wish to thank him for his competent and conscientious copy editing of the book manuscript.

Finally, I would like to mention my greatest debt. Throughout the research and writing of this book Wolfgang Lefèvre untiringly helped me with his encouragement and critical questions. This book is dedicated to him.

Portions of this book have appeared previously in various articles and book chapters, and I thank the editors and presses for their kind permission to use some of this material here. These publications include: "Apothecary-Chemists in Eighteenth-Century Germany," in *New Narratives in Eighteenth Century Chemistry*, edited by Lawrence M. Principe (Dordrecht: Springer, 2007), 97–137; "Die technowissenschaftlichen Laboratorien der Frühen Neuzeit," *NTM* 16, no. 1 (2008): 5–33; "Blending Technical Innovation and Learned Natural Knowledge: The Making of Ethers," in *Materials and Expertise in Early Modern Europe: Between Market and Laboratory*, edited by Ursula Klein und Emma C. Spary (Chicago: University of Chicago Press, 2010), 125–157; "Ein Bergrat, zwei Minister und sechs Lehrende: Versuche der Gründung einer Bergakademie in Berlin um 1770," *NTM* 18, no. 4 (2010): 437–468; "Savant Officials in the Prussian Mining Administration," *Annals of Science* 69, no. 3 (2012): 349–374; "The Prussian Mining Official Alexander von Humboldt," *Annals of Science* 69, no. 1 (2012): 27–68; "Chemical Experts at the Royal Prussian Porcelain Manufactory," *Ambix* 60, no. 2 (2013): 99–121; "Klaproth's Discovery of Uranium," in *Objects of Chemical Inquiry*, edited by Ursula Klein and Carsten Rheinhardt (Sagamore Beach, MA: Watson Publishing International, 2014), 21–46; "Chemical Expertise: Chemistry at the Royal Prussian Porcelain Manufactory," *Osiris* 29 (2014): 262–282; *Nützliches Wissen: Die Erfindung der Technikwissenschaften* (Göttingen: Wallstein, 2016), material from part III; and "Hybrid Experts," in *The Structures of Practical Knowledge*, edited by Matteo Valleriani (Cham: Springer, 2017), 287–306.

INTRODUCTION

The systematic organization of research and expertise promoting technological innovation is characteristic of modern industrial societies. "Knowledge is power," a slogan of the early modern British statesman and philosopher Francis Bacon, is currently on the tip of everyone's tongue. But when exactly did Western societies begin to produce and accumulate technologically relevant intellectual capital? How did this happen? And what ingredients went into the making of this kind of knowledge? There have never been recipes for creativity and novelty, but nuclear power, space flight, superconductive materials, the Internet, intelligent robots, genetically transformed crops, microinvasive surgery, and laser medicine would not exist without the technological sciences and the exact natural sciences. When and how did these types of science come into existence? And what is the relationship between the natural and the technological sciences?

With respect to recent sciences such as nanotechnology, genetic engineering, biotechnology, and robotics, our familiar distinction between the natural sciences on the one hand and engineering or technological sciences on the other seems to collapse. In the framework of these sciences, engineers and scientists cooperate on a regular basis, and it makes little sense to sort out the "scientific" from the "technological" components of these endeavors. Another characteristic feature is their orientation toward industrial, military, medical, and other social demands, although such demands often provide only vague, long-term orientation, leaving engineers and scientists the option to describe their endeavors as "basic" or "fundamental research" rather than "applied science."[1] The Max Planck Society in Germany, for example, defines research carried out at its Institutes for Coal Research and Chemical Energy Conversion as basic research, since it does not aim at immediate application and might be useful only in the long run.

Projects and disciplines like these are often depicted as radically novel types of science, labeled "mode-two science" or "technoscience." Some scholars regard them as the outcome of an "epochal break" within the long tradition of science that led to the dissolution of basic scientific values such as acquisition of knowledge for its own sake, curiosity, truth, and objectivity. Others explain "technoscience," and the "epochal elevation" of technology in its framework, in terms of a culture of "postmodernity" that subsumes science under technology. These scholars assume that in "modernity" pure and disinterested science prevailed.[2] On this view, natural science and technology remained two clearly separated domains well into the twentieth century.

This book questions the argument that scientific and technological inquiry had been separate endeavors before the twentieth century and that recent technoscience is a radically new mode of science. It tackles the following overriding questions: What is the relationship between the more recent technosciences and the older technological sciences? Do interactions, and interdependences, between natural-scientific and technological inquiry have a deeper history? Did institutions that coupled natural-scientific and technological knowledge exist prior to 1900 and perhaps even in the early modern period? And was there any interdependence between the development of the specialized, exact natural sciences and that of the technological sciences? Asking these questions, we will also trace the emergence of the persona who carried out both natural-scientific and technological inquiry and the political and economic milieu in which this was done.

THE BIG PICTURE

The argument for the radical novelty of twentieth-century technoscience relies on the predominant grand narrative in the historiography of science. In a nutshell, this narrative asserts that the early modern predecessor of the modern natural sciences was natural philosophy, supplemented by natural history, medicine, and mixed mathematics. Early modern natural philosophy promoted systematic observation and experimentation with natural objects, seeking to discover universal natural causes and ultimately aiming at a comprehensive, unified understanding of nature. It also established long-lasting values of science such as truth to nature and objectivity.[3] Natural history and medicine were also concerned with *natural* objects— specimens and classes of plants, animals, and minerals, the human body, and diseases. Furthermore, mixed mathematics, comprising fields such as planetary astronomy, geometrical optics, statics, and mathematical geography, is often depicted as a discipline that restructured natural philosophy, moving it away from scholastic Aristotelianism and giving rise to modern mathematical-experimental science.[4]

Because the modern English term "science" means systems of *natural* knowledge, historians of science long took for granted that it is the investigation of nature, and only nature, that constitutes the long history of science.[5] For example, in their introduction to the volume on early modern science in *The Cambridge History of Science*, Katherine Park and Lorraine Daston write that early modern science was the "disciplined inquiry into the phenomena and order of the natural world." Adding immediately that the term "science" is anachronistic with respect to the period before ca. 1850, they point out that natural philosophy "examined change of all kinds, organic and physical" and thus covered a huge area of objects later studied in the natural sciences. In the same volume, Ann Blair writes in her contribution on natural philosophy: "'Natural philosophy' is often used by historians of science as an umbrella term to designate the study of nature before it could easily be identified with what we call 'science' today."[6] In a similar vein, Andrew Cunningham and Perry Williams argue in a well-received essay on the "big picture" in the historiography of science that science, as we know it today, emerged only between ca. 1760 and 1850, and that the predecessors of the modern sciences were "the disciplines of either 'natural history,' or 'mixed mathematics' or especially 'natural philosophy'."[7]

While these historians have highlighted early modern studies of nature, they have marginalized, or even excluded, other important fields that combined studies of natural and technical things—such as chemistry and practical mathematics concerned with fortification, navigation, the construction of optical instruments, land surveying, and so on.[8] Likewise, their big picture marginalizes persons, both in the Renaissance and the centuries afterward, who sought, for example, to understand bird flight in connection with attempts to invent flying machines, or who simultaneously explored the possibility of making chemical remedies and of understanding the nature of chemical transformations. What is more, in the period of the formation of the modern scientific disciplines, approximately between 1750 and 1850, another type of science emerged, comprised of engineering science, agricultural science, and similar technological sciences. In the framework of a grand narrative that focuses on studies of nature, and in particular on natural philosophy, the origins of these kinds of sciences remain obscure.

Historians and philosophers of science have long treated the engineering and technological sciences as the natural sciences' poor relatives, and they have scarcely paid attention to the eighteenth-century sciences of mining, technical chemistry, mechanical engineering, civil engineering, and other "practical" or "useful sciences." "Science" was measured against the standard of "pure science" and epistemic values such as truth to nature, rationality, and objectivity. The historiography of science was thus framed by a concept of science that was only fully articulated by the middle of the nineteenth century and reached its height during the Cold War. Even today, long after the end of that conflict, historians and philosophers of science are rarely concerned with the engineering and technological sciences. The recent debate on "technoscience" also remains fixated on the newest application-oriented research projects at universities, while for the most part neglecting to make the obvious comparison to the tradition of engineering and technological sciences.

There is a second aspect of this grand narrative that invigorates an old premise concerning the early modern knowledge economy: that knowledge was produced, distributed, and accumulated either in the sphere of schools and academies or in that of the arts and crafts, and that there was nothing in between.[9] Although this view grants that the seventeenth-century Scientific Revolution created relationships between the two spheres, it depicts these relationships as mere exchanges of knowledge across intact boundaries. Many publications, both older and more recent, have challenged this view by highlighting converging roles of Renaissance inventors and men of science (Merton), "trading zones" creating long-lasting links between university-educated humanists and workshop-trained practitioners (Long), "shared knowledge" of early modern engineers and men of science (Renn), the "middle epistemology" of eighteenth-century French artillery science (Alder), and so on.[10] Yet the equation *early modern science = natural philosophy* has not lost its canonical status. Asked what the predecessor of the exact natural sciences in the early modern period was, the spontaneous answer of the majority of today's historians of science would certainly be: "natural philosophy." If we turned to historians of technology, asking what the predecessor of modern engineering science was, many would answer: practical-technical knowledge firmly entrenched in artisanal practices.[11] The recent historiography of technology is of little help for studies of the historical development of the technological sciences and their relationship with the exact natural sciences. Some fifty years ago, historians of technology did discuss these issues, but the results of this discussion are largely forgotten today.[12]

The questions raised above are obviously too big to be answered in one book. They should be seen as orientation markers in an ongoing discourse about long-term developments in the history of technology and science.[13] The stories, analyses, and theoretical discussions contained in this book are part of a continued questioning of historians' grand narratives—as many recent studies have done.[14] To stay within our picture of the knowledge economy, we will examine a third sphere situated in between the sphere of local technical knowledge, originating in the craft milieu, and the sphere of natural sciences: that of the so-called practical or useful sciences and the emerging technological sciences. The useful sciences, established in Prussia in the last third of the eighteenth century, offer a unique window into the study of combined technological and scientific knowledge and inquiry before the establishment of the more differentiated disciplinary structure of natural and technological sciences in the second half of the nineteenth century, and hence before a time when daily cooperation between scientists and engineers in research practices became obscured by the institutional differentiation of disciplines. As we will see in chapters 14 and 15, the German technological sciences continued traditions first established in the institutions of useful science.

Zeroing in on useful sciences such as the science of mining, technical chemistry, agricultural science, the science of forestry, and the science of civil engineering (*Bauwissenschaft*), we will also study the persons who helped to institutionalize these sciences and carried out research within their framework. The practitioners of the useful sciences were recognized as men of science (*Naturforscher*) as well as inventive experts. Throughout this book, this hybrid figure is designated as a "scientific-technological expert."

A third, closely related goal of the book is a description of the Prussian state's attempts to organize technical expertise. We will embed the useful and technological sciences as well as their practitioners in the history of the Prussian state bureaucracy, the establishment of technical state departments, and attempts by reform-minded ministers and officials to manage natural resources, improve industry, and compete with industrialized England.[15] Even though these men did not, and could not, foresee the ways in which Prussia would be transformed into an industrialized nation, their beliefs, plans, and activities fostered Prussia's early industrialization.

Throughout the book a terminological distinction is made between "technical knowledge" directly embedded in practice, which is always local knowledge, and "technological knowledge" stemming from the collection and generalization of local technical knowledge and its presentation, for example in the technological books that proliferated from the sixteenth century. This distinction is helpful for our understanding of "useful science," "engineering science," and "technological science" in the eighteenth and nineteenth centuries. It also captures the meaning of the German term *Technologie*, introduced in the second half of the eighteenth century, which denoted school-based, systematic knowledge, on an institutional metalevel so to speak, about technical knowledge, techniques, instruments, materials, and artifacts entrenched in the arts and crafts or in industry.[16]

Of course, today practical engineers' and technicians' technical knowledge is inextricably mixed with school-based technological knowledge and justifies today's comprehensive meaning of the term "technological knowledge." Regarding the terms "engineering science" and "technological science," the latter has a broader meaning than the former, as it comprises sciences such as agricultural science, forest science, nutritional science, technical chemistry, and materials science. In German-speaking countries "technological science" (*Technikwissenschaft*) is a common term, while in Anglophone countries "engineering science" might be more common.

USEFUL SCIENCE

Beginning in the middle of the eighteenth century, the mercantilist states in continental Europe founded distinct institutions for military engineering and the artillery sciences, followed by institutions for the teaching of civil engineering, mining science, agricultural science, and so on. The artillery and navy excepted, in Great Britain and the United States similar institutions were only established in the late nineteenth century. There are excellent studies on the eighteenth-century French engineering and agricultural schools, but comparable studies on Prussia are rare.[17] By studying the Prussian "practical" or "useful sciences," their practitioners, and their political surroundings, this book will illuminate this blind spot.

"Practical science" (*praktische Wissenschaft*) and "useful science" (*nützliche Wissenschaft*) were actors' terms that are largely forgotten today. Unlike early modern scientists' self-advertising assertions that they possessed "useful knowledge," the terms "practical science" and "useful science" originated with reform-minded ministers who promoted scientific and expert knowledge in the Prussian state administration.[18] The useful sciences—ranging from mining science, technical chemistry, and the science of civil engineering to agricultural science and forest science—were established at new kinds of institutions, such as mining academies, schools of agriculture, schools of forestry, the Academy of Civil Engineering and Architecture (Bauakademie) in Berlin, and in less elaborated forms even at the Royal Prussian Academy of Sciences and at the Prussian universities. They were tailored to the attempts of reform-minded ministers and officials to solve technical problems and improve industrial technology. Yet "useful science" did not signify applied natural science. Rather, the useful sciences interconnected scientific, technological, and practical-technical knowledge and inquiry. Hence, they represented a middle ground that bridged the world of the natural and mathematical sciences and the world of agriculture, forestry, and industry.

Exploring the horizon of possible transformations in the material world, drawing on both nature and the arts and crafts, practitioners of the useful sciences used natural knowledge to study the workings of mining machines, the structure of ore deposits, transformations of raw materials into refined substances, the improvement of agriculture and forestry, and many other technical items. Conversely, they used insights gained from technological studies for the advancement of natural knowledge. With respect to teaching, useful science meant the combination of theory and practice, that is, teaching in the classroom as well as practical training. In the mining academies at Freiberg and Schemnitz, for example, students trained in the mines in addition to attending classes. In a similar way, students of the Academy of Civil Engineering and Architecture, the War School (Allgemeine Kriegsschule; later Preußische Kriegsakademie), and the United Artillery and Engineering School (Vereinigte Artillerie- und Ingenieurschule) not only attended lectures but also received practical training, for example, in surveying and civil engineering. The Industrial Institute (Gewerbeinstitut) continued this tradition. Education by lectures was supplemented by training in workshops and laboratories as well as in private factories.

Seen from a later perspective, the institutionalization of useful science was a major step toward the formation of the specialized engineering and technological sciences and the exact natural sciences. The latter will be detailed with respect to analytical chemistry, metallurgical chemistry, mineralogy, geology, and botany. These kinds of natural-scientific disciplines, or subdisciplines, were significantly promoted in the context of the eighteenth-century useful sciences and the nineteenth-century technological sciences.[19] Similar processes have been observed with respect to the eighteenth-century engineering and artillery sciences and parts of theoretical mechanics such as ballistics and statics.[20]

The useful sciences and the specialized natural sciences were pushed forward by reform-minded ministers and officials, not by the Prussian kings and the court. Whereas the ministers organizing the useful sciences concerned with mining, forestry, and other civil sectors struggled with a lack of funds, the Prussian kings were ready to invest only in teaching institutions that were also of benefit to the military. The Medical-Surgical College in Berlin (Collegium medico-chirurgicum), founded in December 1723, was one such institution, which was intended to be primarily a training school for military surgeons. It possessed an anatomical theater, had access to the Royal Botanical Garden and the laboratories of the Royal Court Apothecary Shop, and its teachers offered not only lectures but also experimental and practical courses.[21] On account of its unconventional empirical methods, it soon became a center of attraction for future physicians, apothecaries, and a variety of other men interested in the sciences. As a consequence, in 1795 King Friedrich Wilhelm II (reigned 1786–1797) approved the establishment of a new school for military surgeons, the Chirurgische Pépinière (later Friedrich-Wilhelms-Institut), which was placed under the supervision of a military director and charged exclusively with "the training of competent company surgeons for the field regiments in perpetuity."[22] At the end of the 1780s, the king additionally approved the foundation of a school for veterinarians and horse doctors, which served the interests of the cavalry. The Royal Veterinary School, which opened its doors in 1791, was housed in a splendid building designed by Carl Gotthard Langhans in the Spandauer Vorstadt.[23] In addition to these medical-military training facilities, there were some schools fully under the direction of the military, such as the Engineering Academy in Potsdam and the Artillery Academy, founded in 1791 by the artillery officer and mathematician General Georg Friedrich L. von Tempelhoff. However, unlike the famous French artillery and engineering schools, their eighteenth-century Prussian counterparts were small, isolated institutions with little impact on the public discourse on useful science. This changed only in the nineteenth century, in connection with the Prussian army's modernization, led by General Gerhard von Scharnhorst (1755–1813), and Scharnhorst's foundation of the War School in 1810. The reformers in the War School and the United Artillery and Engineering School, which split off in 1816, fostered the military's participation in public debates on scientific-technological education and Prussia's industrialization.[24]

SCIENTIFIC-TECHNOLOGICAL EXPERTS

The knowledge economy that underlies the grand narrative in the historiography of science outlined above ascribes knowledge either to scholars or to artisans and craftsmen. By contrast, we will see that there was a third persona in between the natural philosopher and the traditional craftsman. In fact, there was a whole spectrum of figures in between these two extremes, ranging from engineers, architects, surgeons, itinerant inventors, and other "technical experts" to men recognized as both technical experts and men of science. The hybrid figure of "scientific-technological expert" is of particular interest with respect to the institutionalization of the "useful sciences" and its nineteenth-century successors, the technological sciences and the specialized, exact natural sciences.

Scientific-technological experts were the most important practitioners of the useful sciences. A major goal of this book is the dense description of their knowledge and activities. Scientific-technological experts elaborated curricula for the new technological schools, were themselves active as teachers, and carried out experiments and work of invention. Further, they reflected on the conception of useful science, writing programmatic texts on this subject. Almost all were also state advisors or officials. Men like Franz Carl Achard, Martin Heinrich Klaproth, Carl Abrahm Gerhard, and the young mining master Alexander von Humboldt exemplify a multitude of scientific-technological experts in late eighteenth-century and nineteenth-century Prussia who do not fit into the conventional images of the romantic German *Naturforscher* or, alternatively, the untutored technician.[25] Rather, these men brought about a close interaction between natural-scientific and technological inquiry, and carried out practical technical work in the framework of state-managed mining, civil engineering, forestry, and agriculture. We will discuss the roles played by scientific-technological experts from a broader historical perspective, and compare the concept of scientific-technological experts with the older scholar-and-craftsman thesis; Charles Gillispie's view of the relationship between savants and technicians in eighteenth-century France; and Steven Shapin's distinction between gentleman scientists and invisible technicians in early modern England. In the second half of the nineteenth century, the "engineer-scientist" was the most important representative of the figure of scientific-technological expert.[26]

PRUSSIA'S INDUSTRIALIZATION

The period covered in this book, 1750–1850, was the time of the first Industrial Revolution in Europe. It was a period of rapid technical innovation and profound economic and social transformations wrought by the expansion of industrial capitalism. The complexity of the process has long been a historiographical challenge, manifested not least by the divergent approaches historians have chosen to unravel its different threads.[27] We will look at the early industrialization of Prussia through the prism of industrial policy, especially the state's attempt to improve industrial technology and organize technical expertise.

The grand narrative in the historiography of science addressed above has also affected our view of the Industrial Revolution. It is often assumed that the foundation of the synthetic dye industry and the electrical industry in the second Industrial Revolution of the last third of the nineteenth century was a turning point in the relationship between science, technology, and industry, as these new industries implemented advanced technological as well as chemical and physical knowledge. By contrast, the drivers of the first Industrial Revolution in eighteenth-century England have long been depicted as practical tinkerers, without any technological and scientific education. Historians such as Margaret C. Jacob, Joel Mokyr, and Larry Stewart have challenged the latter view, arguing that the early modern scientific institutions and the Enlightenment movement created an intellectual climate that fostered the production of knowledge, including scientific knowledge relevant to technology and industry.[28] While Jacob and Mokyr initially understood "science" to denote Newtonian theory, their later concept of science is broader and includes empirical knowledge as well. More recent research by Peter Jones, David Miller, and Leslie Tomory has detailed the ways in which entrepreneurs and engineers such as Matthew Boulton, Josiah Wedgewood, and James Watt implemented scientific materials and instruments in their work of invention and merged scientific with technological knowledge.[29] Peter Jones has introduced the term "savant-fabricant" for these pioneers of the English industrialization.[30]

Studies like these have shown that the proliferation of scientific and technological publications, public lectures, and communication at meetings of the Lunar Society and other scientific associations were constitutive of a British "industrial enlightenment" (Mokyr). In the mercantilist and absolutist states of the European continent similar processes were happening, wherein the state, however, had a substantially more active role. In France, Austria-Hungary, Saxony, and in Prussia as well, technical experts with artisanal backgrounds entered into public service. Likewise, men of science with technical expertise became advisers in the framework of state-directed technical projects. We will illuminate Prussia's industrial policy and promotion of expertise from the 1740s and during the reign of Friedrich II, when mercantilist policy was reinforced, until the 1850s, when the state began to withdraw from the economy and the first wave of the mechanization of private factories had already peaked.[31]

INDUSTRIAL POLICY

During the reign of Friedrich II (1740–1786, known to the English-speaking world as Frederick the Great), the central Prussian governmental and administrative agency in Berlin, the General Directory of War and Finances, was slowly transformed into a functional state bureaucracy concerned with a broad range of public affairs, including manufacture.[32] Reform-minded ministers and officials began to define a new role for the state comparable to the French mercantilist policy that had been developing since the era of

Colbert. Whereas in the old regime, the main functions of the state authorities had been tax collection and auxiliary services for the army, the General Directory now had to become a powerful instrument for the improvement of industry and commerce. Thus, in 1740 a Department of Trade and Commerce was created, which was the first technical state department that supplemented the four main departments of the General Directory responsible for distinct provinces. As a consequence, natural and technological knowledge were recognized for the first time as indispensable prerequisites of policy.

In the aftermath of the Seven Years' War (1756–1763), this trend was reinforced. An economic depression made the industrial hegemony of Great Britain more tangible, which created favorable conditions for further reforms of the state administration. Reform-minded ministers and leading officials thus convinced Friedrich II to strengthen the technical departments of the state bureaucracy. In May 1768, the General Directory established a Department of Mining and Smelting Works (Bergwerks- und Hüttendepartment). This was followed by the foundation of a Department of Forestry (Forstdepartment) in January 1770 and a General Department of Building or Civil Engineering (Oberbaudepartment) in June 1770. The reform-minded ministers agreed that the new type of state official serving in these technical departments must be an expert in his field, possessing practical-technical, technological, and scientific knowledge, as well as respect for the values of science. In line with this ideal was the concept of useful science, which stabilized these values and organized technologically relevant knowledge.

Through the annexation of Silesia during the three Silesian Wars (1740–1763), Prussia gained access to rich iron resources. The iron industry became an important military and economic factor, which triggered Prussia's industrialization. Expert officials in the Mining and Smelting Works Department managed iron production, undertook experiments to improve the quality of iron and steel, and organized the teaching of the mining sciences. Like officials working in the Department of Forestry, these officials also searched for ways to save wood, because the shortage of wood was perceived as a factor limiting the mining industry. For example, they examined the possibility of replacing wood with coal.[33] The late eighteenth century also saw the introduction of steam engines and coke blast furnaces in state-directed Silesian mining. Consequently, between 1800 and 1835 Prussian raw iron production increased by 135 percent and steel production by 100 percent.[34]

With the beginning of Prussian industrialization in the nineteenth century the state's technical endeavors expanded significantly. This was facilitated by a reorganization of the state administration in 1808 that transformed the General Directory into the so-called State Ministry (Staatsministerium), which was strictly organized according to political, economic, and cultural functions. Mining, a key factor for machine construction and hence for the mechanization of manufacture and transportation, remained in the hands of the state until the 1860s. The number of state-owned factories increased through the activities of state agencies such as the Overseas Trading Corporation (Seehandlung).[35] New state organizations such as the Technical Deputation undertook measures to smuggle machines from England, foster Prussia's engineering industry, and replace handwork by machines in private industry. The state also expanded its activities in the transportation sector from the building of canals, bridges, and streets to the construction of railways. Moreover, state-organized schools such as the Academy of Civil Engineering and Architecture and the Industrial Institute became centers for the education and training of engineers and technical experts (*Techniker*). In addition, the state undertook a number of economically relevant legislative measures such as tax reforms, the abolition of guild compulsion and introduction of freedom of trade in 1810, the introduction of the Prussian patent law in 1815, and the founding of the German customs union (Deutscher Zollverein) in 1833/34.

The stories and analyses presented in this book speak for the view that there was a slow, continuous process of industrialization in Prussia, which began in the late eighteenth century and was, in its initial phase, fostered by reform-minded ministers and expert officials—not least through the organization of technical expertise. The contrary view that Prussia's industrialization abruptly "took off" in the 1850s has long been questioned in the literature but is still widespread.[36] Unlike the present study, which concentrates on technological change, the latter view hinges on the emphasis of economic factors in industrialization and the expansion of industrial capitalism. The related historical studies typically detail the role played by private entrepreneurship in Prussia's industrialization. The two approaches—focusing on technology or on economic factors—by no means exclude each other. Rather, they ought to be seen as complementary approaches. This presupposes reflection on the ideological questions involved here. The most contested issue in this context concerns the role played by the state in Prussia's industrialization.[37] Even though the intellectual discourse on cameralism was replaced in the nineteenth century by laissez-faire liberalism, the Prussian state long continued to intervene in technology and industry.[38] We will not endeavor to solve the question of whether this was good or bad for Prussia's industrialization, or what would have happened if the state had withdrawn earlier. Instead, we scrutinize the consequences of the state's activities for the development of technical expertise and of the sciences that linked natural and technological knowledge and inquiry.

STRUCTURE OF THE BOOK

Part I of the book studies the Prussian state's recruitment of experts for technical projects and manufacturing from the 1750s until the 1790s, that is, in a period when schools dedicated to the education and training of experts were still rare. Starting in the 1740s, the General Directory undertook numerous technical projects such as land surveys, drainage of marshland, construction of canals, investigation of new construction materials, and installation of lightning rods. In order to carry out these endeavors, it recruited members of the Royal Prussian Academy of Sciences as well as artisans considered to be outstanding experts. Chapter 1 illuminates the technical activities and inventions of members of the Royal Prussian Academy of Sciences who worked on such state projects. These men stood with one leg in the world of the sciences and the other in state-organized technical projects and manufacturing. They exemplify the figure of scientific-technological expert highlighted in this book.

The apothecary trade, studied in chapters 2 and 3, was an artisanal sector that especially fostered expertise on materials and material transformation. It thus prepared the ground for more advanced chemical inquiry and especially for pharmaceutical chemistry. The Prussian ministers often recruited technical officials and advisers from this particular artisanal field. Thus, in the 1780s the chemist and apothecary Martin Heinrich Klaproth became a scientific-technological adviser of Minister Friedrich Anton von Heinitz, responsible for Prussian mining. Chapter 2 illuminates Klaproth's scientific and practical activities. Putting his discovery of uranium into the context of Heinitz's industrial policy, it provides an example of the interdependence of discovery and invention. The chapter also shows how Heinitz's policy promoted analytical and mineralogical chemistry. Chapter 3 examines the relationship between the eighteenth-century apothecary trade and chemistry from a more general point of view, studying their shared material culture and the social incentives that triggered apothecaries' shifts from pharmaceutical production to chemical inquiry.

The fourth chapter turns to another production site: the Royal Prussian Porcelain Manufactory in Berlin. Whereas the apothecary trade was a private artisanal sector dominated by small workshops, the Royal Prussian Porcelain Manufactory was a large establishment employing several hundred workers. The chapter provides insight into the role played by the

Manufactory's technical experts in a system of division of labor. It also illuminates the organization, established by Heinitz in the late 1780s, of combined practical training and scientific education of the Manufactory's technical experts. Chapter 5 summarizes the characteristics of the figure of the technical expert and sheds light on its history.

Part II moves on to state-directed mining, mining science, and mining academies in the second half of the eighteenth century. Mining was a driving factor in the industrialization of Europe, yielding iron and steel for the construction of machines, transportation devices, and factory halls. The centuries-old silver mining industry, in particular, belonged to the high-tech industry of the time. It was a hothouse for technical expertise and cross-fertilizations of natural-scientific and technological inquiry, and in the German states it was the first civil sector to develop a demand for useful science. The science of mining, elaborated and taught at mining academies in the old silver-mining regions, became a model for the institutionalization of the useful sciences more broadly, especially with respect to the combination of theory and practice. Chapter 6 provides an overview of silver mining in Saxony, its system of division of labor, knowledge economy, and different types of mining experts and officials. Particular emphasis is placed on the Freiberg Mining Academy in Saxony, founded in 1765, which the reform-minded Prussian ministers and officials regarded as a milestone in the institutionalization of mining science and useful science.

Chapter 7 provides insights into the activities and goals of the Department of Mining and Smelting Works in Berlin. Taking the mineralogist and mining counselor Carl Abraham Gerhard as an example, it describes the roles of Prussian mining officials and the ways in which they sought to improve mining. Gerhard was a scientific-technological expert who bridged the worlds of science and mining and contributed to the development of mining science.

Chapter 8 describes his chemical-technological experiments on mineralogical materials, carried out in the laboratory of the Mining and Smelting Works Department that Gerhard had set up himself. Both the laboratory and Gerhard's experiments illuminate the interdependence, and local convergence, of chemical and technological inquiry in the context of mining and mining science.

Chapter 9 reconstructs Gerhard's attempt to establish a mining academy in Berlin, which failed not least because Friedrich II was unwilling to finance the new institution. As a replacement, the mining administration organized a modest lecture series. Even so, this lecture series became a framework for training Prussia's lower, technical mining officials.

Chapter 10 refines the picture of interconnected scientific and techno-logical inquiry in the field of mining, studying the professional technical activities, work of invention, and scientific experiments of the young min-ing master Alexander von Humboldt.

While part II has a narrative structure and describes practices of mining science along with its practitioners and institutions, part III tackles ques-tions concerning "useful science" in a more general way. Chapters 11 and 12 analyze unpublished reports and a programmatic text concerned with mining science and the science of saltworks, which provide unique insight into the historical actors' understanding of "useful science." We will see that the late eighteenth-century sciences of mining and saltworks were regarded as mixed technological and scientific disciplines, established for teaching as well as research. The old question concerning the "how" of the relevance of early modern natural sciences for technology and industry is thus illuminated from a new perspective. Further, chapter 11 addresses the pragmatic epistemology developed in the framework of mining science along with new epistemic values such as reliable knowledge. These analyses are supplemented at the end of chapter 12 by a brief overview of the his-torical roots of useful science and a discussion of its relationship with the university disciplines of cameral science and "technology."

Chapter 13 takes a closer look at the most important practitioner of use-ful science: the scientific-technological expert. From the last third of the eighteenth century until the middle of the nineteenth century, scientific-technological experts played a crucial role for the institutionalization of the useful sciences, their teaching, their refinement through research, and their further transformation into the technological and the specialized, exact sci-ences. The chapter describes the characteristic features of this hybrid figure and discusses alternative interpretations.

Part IV turns to new scientific institutions and events in the nineteenth century, when Prussia's industrialization gathered momentum and useful science was slowly transformed into technological and specialized natural sciences. Chapter 14 studies the significance of the foundation of the Uni-versity of Berlin in 1810 for the teaching of useful science. Showing that the university in its early years integrated parts of the mining administra-tion's lecture series in its curriculum and furthermore created professor-ships for technology, agricultural science, and forest science, it questions the common view that the young University of Berlin was a modern research university marked by the neohumanist ideal of *Bildung* and dedi-cated exclusively to pure science. The Academy of Civil Engineering and Architecture (Bauakademie) in Berlin, studied in chapter 15, was an early citadel of the teaching of architecture and civil engineering, or what the Germans designated as "the science of civil engineering" (*Bauwissenschaft*). It addressed future officials of the Prussian state's General Building Department, ranging from land surveyors to leading officials such as building counselors. The chapter details reformers' struggles to secure support for that institution immediately before its foundation in 1799, and provides an overview of its complex history until 1879, when it was unified with the Industrial Institute into the Technical University of Berlin (Königliche Technische Hochschule zu Berlin). Like the Mining Academy of Freiberg, the Academy of Civil Engineering and Architecture offered courses on science and technology and connected theory with practice, that is, classroom teaching with practical courses in land surveying and work at the state's construction sites.

Chapter 16 turns to two new institutions established for the promotion of Prussia's private industry: the Technical Deputation and the Industrial Institute, founded in 1821 by the influential Prussian official Christian Peter Wilhelm Beuth. Beuth was the strategic head of a new industrial policy that aimed at introducing machines in Prussia's private industry and organized the education and training of engineers and *Techniker* (technical experts) for that purpose. He systematically organized the transfer of technological knowledge and machines from England, created a platform for their investigation and replication, and further organized the donation of replicated machines to private entrepreneurs. His most sophisticated strategy consisted of rendering former students of his Industrial Institute personal agents for the introduction of machines in private factories, which guaranteed both his students' employment and the mechanization of industry. In the framework of this system, the fluctuating contours of the technical expert gave way to the well-defined figure of the nineteenth-century engineer (*Ingenieur*).[39]

The final chapter (17) is the most theoretical one. Summarizing previous analyses of useful sciences, it brings to the fore the main epistemological features of this new type of science. The chapter goes on to compare eighteenth-century useful science, nineteenth-century technological science, and recent technoscience. The comparison leads to the conclusion that eighteenth-century useful science and nineteenth-century engineering and technological science ought to be understood as early modes of technoscience. The chapter ends with considerations concerning the big picture of the history of science and technology.

TECHNICAL EXPERTS AND INNOVATION IN PRUSSIA

1 TECHNICAL PROJECTS OF THE ROYAL PRUSSIAN ACADEMY OF SCIENCES

In the eighteenth century, distinct schools for the education and training of technical experts were still a rarity in Prussia. Only after the Seven Years' War did the Prussian state intensify its efforts in this area. Before this time, the General Directory and the king had occasionally financed projects of itinerant inventors or asked foreign experts for technical advice. Within the military, several schools had been established specifically for the training of engineers, artillery officers, and military surgeons. However, for its technical enterprises in the civil sectors, the General Directory long relied on ad hoc recruitment of experts.

The Royal Prussian Academy of Sciences was one of the most important institutions the Prussian state administration turned to when it was in need of experts for the evaluation and organization of its technical projects. Practical utility of knowledge at the service of the common good had already been emphasized by the Academy's predecessor institution, the Society of the Sciences (Societät der Wissenschaften), founded in 1700 according to a plan by Gottfried Wilhelm Leibniz (1646–1716). The Society's charter stated that it should organize inventions and improve works of the mechanical arts, and to this end it should, as Leibniz had pointed out, unite "learned people, engineers, and artists" under one roof.[1] In 1743, under the reign of Friedrich II, the Society was renamed the Académie des Sciences et Belles Lettres and divided into four classes: the physical, mathematical, philosophical, and historical-philological classes.[2] Useful knowledge was mainly the concern of the physical and mathematical classes. Their members contributed to the state's technical projects, gave advice to ministers, and worked on inventions themselves.

In the middle of the eighteenth century, the members of the mathematical class participated in the state-organized drainage of the marshlands of the Oder, Warthe, and Netze rivers. The Oder valley between Oderberg and Küstrin was a large marshland, traversed by tributaries. Around 1750,

Leonhard Euler (1707–1783) participated in the drainage of this part of the Oder marshlands, through which some sixty thousand hectares of land were reclaimed and subsequently settled with villages.[3] As the measures required extensive land surveys, the state employed Euler for their planning and management, as Euler had already acquired practical experience through his involvement in the reconstruction of the Oder-Havel-Finow canal between 1744 and 1746.

FIGURE 1.1
The Marstall building, with the observatory of the Royal Prussian Academy of Sciences, in the Letzte Straße in 1745.
Source: Buddensieg, Düwell, and Sembach 1987, 27.

Euler's successors, Joseph-Louis Lagrange (1736–1813) and Johann Heinrich Lambert (1728–1777), also took part in land improvement projects. As an honorary member of the General Building Department, Lambert participated in numerous civil engineering projects.[4] After the Seven Years' War, the Prussian state organized a *rétablissement*, renovating buildings and rebuilding villages and towns destroyed in the war.[5] This policy increased the need for building materials and involved challenges such as fire security and structural stability. The shortage of wood was another problem that stimulated the search for new construction materials and building techniques such as solid building.[6] Hence, in 1777 the Academy formed a committee for the investigation of building materials such as limestone and mortar. Initial participants included Lambert, the members of the physical class Andreas Sigismund Marggraf (1709–1782), Franz Carl Achard (1753–1821), and Carl Abraham Gerhard (1738–1821), the philosopher Johann Georg Sulzer (1720–1779), and, after Lambert's death in September 1777, the mathematician and building councilor Johann Karl Gottlieb Schulze (1749–1790). The committee worked closely with a master mason named Welz, in whose house the materials were investigated and experimental arches and pillars were erected. The committee accompanied these experiments, organized the protocols, and met regularly for consultations.[7]

Among the members of the mathematical class, the artillery officer Georg Friedrich L. von Tempelhoff (1737–1807), who became a member of the Academy shortly after the death of Friedrich II in 1786, forged a personal link between the Academy and the military. Tempelhoff first organized a local artillery school for Prussian artillery officers and engineers, which was followed in 1791 by the foundation of the Artillery Academy in Berlin. Around this time, he had become an influential person and was promoted to the rank of general. By contrast, during the entire reign of Friedrich II, members of the artillery and the corps of engineers received little support from the king.[8] Further, as military secrecy also prevented the greater intellectual outreach of the artillery schools, Tempelhoff was the only artillery officer who acquired an international scientific reputation, mainly through the publication of a Newtonian treatise on ballistics, *Le Bombardier Prussien* (1781). In Prussia, his mathematical and physical work became more widely known through his students, in particular those who left the military. One such student was the building councilor and director of the Academy of Civil Engineering and Architecture, Johann Albert Eytelwein (1764–1848).[9] Within the military, Tempelhoff supported political and technological-scientific reformers such as Gerhard von Scharnhorst, who founded the War School (Allgemeine Kriegsschule) in 1810.

Advising ministers was an obligation of all academicians, but the members of the physical and mathematical classes were naturally the most active persons in this respect. In connection with consultation, the academicians often undertook technical and experimental work both individually and as a collective. For example, in the spring of 1780, Johann Gottlieb Gleditsch (1714–1786), Marggraf, and Achard, who were all members of the physical class, coauthored an evaluation of a petition of a merchant named Carl Ludwig Müncheberg, who wished to set up a workshop for a new sort of vinegar. The evaluation was preceded by experiments and chemical analyses of the vinegar conducted by Gleditsch and Achard, the results of which were summarized by Marggraf, who was the director of the physical class.[10] In some cases, members of different classes cooperated in an evaluation. For example, in the summer of 1777, the mathematical class members Lambert, Nicolas de Beguelin (1714–1789), and Jean Castillon (1708–1791), the physical class members Gerhard and Achard, and the philosophical class member Sulzer coauthored a report on the planned installation of a lightning rod on the roof of the artillery's powder depot at Schlesisches Tor. They formed a commission, wrote individual evaluations that they reciprocally commented on, and then cowrote the report. The practical part of the project was Gerhard's responsibility, who took over the "supervision and direction of this affair" as well as the "direction of the workers."[11]

Several members of the physical class also were state officials who forged long-term links between the Academy and the ministries. Among these were the chemist Sigismund Friedrich Hermbstaedt (1760–1833) and the mineralogists Johann Gottlob Lehmann (1719–1767), Johann Jacob Ferber (1743–1790), Dietrich Ludwig Gustav Karsten (1768–1810), Gerhard, and Alexander von Humboldt (1769–1859).[12] Hermbstaedt was a major figure in the state's attempt to teach the useful sciences and foster the mechanization of the Prussian textile industry in the first half of the nineteenth century. Gerhard was a long-established mining expert, experimenter, and teacher of the science of mining, who organized lectures under the authority of the Berlin mining administration. Humboldt, who became a member of the Academy in 1800, was actively involved in mining technology and in related experimentation from 1792 until 1796/97. In addition to this group of officials, Martin Heinrich Klaproth (1743–1817) was a major consultant of the powerful Minister Friedrich Anton von Heinitz (1725–1802).[13] All of these men exemplify the persona of the scientific-technological expert.

PROJECTS ON FORESTRY AND CULTIVATION OF PLANTS

Among the technical projects carried out by the members of the Academy of Sciences, the cultivation of silkworms and mulberry trees was the largest and longest-lasting enterprise. But especially the members of the physical class undertook numerous smaller projects as well. Examples are the invention of new pigments for the Royal Prussian Porcelain Manufactory by the chemists Achard and Klaproth, described in chapter 4; the cultivation of useful plants and investigations in forestry and agronomy by the biologists Gleditsch, Achard, Carl Ludwig Willdenow (1765–1812), and Albrecht Daniel Thaer (1752–1828); the extraction of sugar from maple syrup by Hermbstaedt; and Achard's project on the extraction of beet sugar, which we will return to shortly.

Johann Gottlieb Gleditsch applied his botanical knowledge to forestry and the science of forestry, which he helped to establish. Wood was the most important construction material and energy source in the eighteenth century, which some historians have designated "the epoch of wood."[14] The shortage of wood became an important topic in the German cameralist discourse and a crystallization point for the rising field of forestry. In 1770, a Department of Forestry was founded in the Prussian state administration, which was responsible for the management of forests, cultivation of new kinds of timber, and organization of their "sustained use" (*nachhaltige Nutzung*). The concept of "sustained use," introduced in 1713 by the Saxon mining official Hannß Carl von Carlowitz (1665–1714), epitomizes the authorities' awareness that industrial progress could be secured only if overexploitation (*Raubbau*) of nature was avoided.[15] This was a climate that stimulated inquiry in forestry and the formation of forest science.

Gleditsch had studied medicine, mathematics, and philosophy at the University of Leipzig, where he gained a doctorate in medicine. In 1744, he was admitted to the Royal Prussian Academy of Sciences. Two years later he was appointed professor of botany at the Medical-Surgical College of Berlin. In the fall of 1770, shortly after the foundation of the Department of Forestry, he began to teach the science of forestry in the context of the newly established series of lectures on the useful sciences, organized by the mining administration. Relying on mining administration records, he also organized excursions to forests in order to demonstrate problems of forestry.[16] In 1774–1775, he published a two-volume textbook on forestry, *Systematic Introduction into the Newer Forestry* (*Systematische Einleitung in die neuere ... Forstwirtschaft*).[17]

Since 1746, Gleditsch had also been the director of the Royal Botanical Garden in Berlin, which was subordinate to the Academy of Sciences. This enabled him to include practical courses in his botanical teaching, as well as to carry out botanical experiments and to cultivate useful plants. The Academy of Sciences had long used the Royal Botanical Garden for economic purposes and had cultivated medicinal herbs, mulberry trees, and exotic fruit trees, to which Gleditsch added a new nursery. His most famous botanical experiments studied the sexual reproduction of plants.

In the 1810s, another renowned botanist of the Academy of Sciences, Carl Ludwig Willdenow, transformed the Royal Botanical Garden into a site of systematic research on Linnaean botany. But he also undertook numerous useful projects. Willdenow, who was a nephew of Gleditsch, had first completed a pharmaceutical apprenticeship in his father's apothecary shop in Berlin. He then studied medicine at the University of Halle for two years and attended Wiegleb's chemical-pharmaceutical boarding school in Langensalza for another year. In 1789, he became a doctor of medicine and started to work in his father's apothecary shop. He ran his father's shop until 1798, when he was appointed professor of botany at the Medical-Surgical College. While working as an apothecary, he published widely on botanical subjects, first in Latin on Berlin's flora and then in the vernacular on herbalism and arboriculture. He also extended his research interests to include physiology and plant geography.

Like all members of the Academy's physical class, Willdenow evaluated numerous practical projects under the authority of the Academy or the state administration. For example, in spring of 1801 the Academy's directors asked him to examine the use of *Lupinus albus* as a substitute for tea and coffee. The latter had been suggested by a woman, named Werneburg, who had also provided the beans of the plant for cultivation in the Academy's garden. It was one of the rare cases where a woman approached the Academy, which otherwise was a purely male domain. Willdenow made a "self-experiment" with the plant. "I have made a trial with the plant's tea and peels," he reported, "and I must confess that [the tests] of the two were quite positive. Therefore, I think it would be worthwhile to make further experiments with this and other kinds of *Lupinus*."[18] In the summer of 1802, Willdenow began to perform crossing experiments with different kinds of potatoes, which pursued the dual goal of studying vegetable reproduction and cultivating new sorts of potatoes. Not all academicians were pleased with his use of the Academy's library as a storage room for the potatoes in the following winter. Similar experiments followed in the next years, first on the cultivation of sunflowers for the production of cooking oil, and then on the germination of grain that had been treated with iron vitriol.[19]

ACHARD'S INVENTION OF BEET SUGAR

The most inventive member of the Royal Prussian Academy of Sciences in the last third of the eighteenth century was undoubtedly Franz Carl Achard. Neither in his private nor in his professional life was Achard a man who would have easily adapted himself to social norms and rules. He provoked scandals that the Academy had to extract him from, and he repeatedly went deep into debt through his endeavors.[20] The tireless man undertook numerous scientific projects, comprising fields as diverse as the studies of gases, meteorological measurements, chemical analyses of gems, chemical studies of acids and salts, electrical experiments, botany, and plant physiology. Of his technical projects, undertaken for the most part under the authority of the Academy, we have already mentioned the installation of lightning rods, the examination of building materials, and the invention of pigments for the Royal Prussian Porcelain Manufactory. He also worked on the production of alloys—a continuation of experiments carried out by his friend Carl Abraham Gerhard (see chapter 8)—and on artificial mineral waters, ballooning, and several horticultural and agricultural projects. But his most famous project was on the invention of beet sugar, which will be described in the following.

Achard came from a wealthy and influential Huguenot family. He received a private education and was in contact with Berlin's intellectual elite from early on. Two of his uncles, who were responsible for his education, were philosophers and members of the Royal Prussian Academy of Sciences. In 1772, he began to attend the lecture series on the useful sciences that the Mining and Smelting Works Department had started shortly before. Two years later, he was already known as a man of science and owner of a collection of expensive physical apparatus. In June 1776, at the age of twenty-three, he became a regular member of the Academy of Sciences and began to experiment in its laboratory under the direction of Marggraf. In 1774, after Marggraf had suffered a stroke, he became Marggraf's assistant. Working in this position, he acquired the experimental skills he later needed for his chemical analyses and inventive work on beet sugar. Marggraf had been an apothecary, working as the administrator of his father's apothecary shop in Berlin from 1735 until 1752. In 1747, while analyzing vegetable substances, he discovered that domestic beet varieties contained a kind of sugar that was identical to the expensive, imported cane sugar.[21] He immediately pointed out the potential utility of his discovery, but did not undertake any further experiments on the large-scale extraction of sugar from beets. Instead, he continued his project on the

FIGURE 1.2
Portrait of Franz Carl Achard.
Source: Deutsches Museum München.

chemical analysis of vegetable substances. All of these experiments were performed in his own pharmaceutical laboratory.

Already in his first years as an academician, Achard had studied the connection between the composition of soil and plant growth, and cultivated useful plants in the Academy's small garden in the Letzte Straße. Field experiments on the cultivation of exotic tobacco varieties followed in the fall of 1780, first on Virginia tobacco and then on Asiatic tobaccos. Because these experiments fitted into Prussia's mercantilist policy, Friedrich II provided an annual stipend of five hundred thalers. Achard continued these experiments until at least 1786, although in the early 1780s his main interests had turned to beets and beet sugar.

In August of 1782, Achard succeeded his recently deceased mentor Marggraf as the director of the laboratory and of the Academy's physical class. He thereby joined the ranks of the most influential academy members, with a laboratory and a cabinet of instruments at his disposal. Around the same time, he started his sugar beet project, which comprised two phases: first, the cultivation of sugar beets, and second, the extraction of sugar from beet juice on a large technical scale. The entire project took eighteen years.

As a first step, Achard bought a manor in Kaulsdorf, a village east of Berlin, and started to tackle the first challenge of his project, the cultivation of sugar beets with a high sugar content.[22] In the following three years he experimented with numerous sorts of sugar beets, selecting for the highest sugar content. In addition, he carried out experiments on manure, performed chemical analyses of soil, and tested new methods of tillage. After a large part of his manor had been destroyed in a fire in September 1783, he sold it in 1785. A period of five years followed in which he only worked theoretically on agricultural topics. At the beginning of 1790, he recommenced his experiments on sugar beet cultivation at a newly acquired manor in Französisch Buchholz, a Huguenot settlement east of Berlin. He performed these experiments through the end of 1798. During this eight-year period, he also studied a broad range of other agricultural subjects, about which he reported in publications, at the weekly meetings of the Academy of Sciences, and at the meetings of the Economic Society of Potsdam (Märkische Ökonomische Gesellschaft zu Potsdam).[23] In accordance with the Economic Society's goals, he aimed for knowledge-based and experimentally developed agricultural innovation. Thus, he also began to experiment with English ryegrass as cow fodder, which was supposed to increase the quantity of milk and its fat content. Further, he extended his plant breeding experiments, which served both botanical and practical goals. In 1796, he published a catalogue in which he described 2,311 kinds of plants he had cultivated in his garden and fields, and also offered plant seeds for sale.[24]

TRANSFORMATION OF THE ACADEMIC LABORATORY INTO A SUGAR FACTORY

Achard used private money to finance the cultivation of sugar beets, for which he went deep into debt. By contrast, his subsequent experiments on the production of sugar from beet juice were mainly financed by the king and the Academy of Sciences. In November 1798, Achard began the first experiments on large-scale production of beet sugar in the Berlinische

Zucker-Siederei-Companie. Three positive evaluation reports, one of which was penned by Klaproth, recommended the continuation of the experiments. In the meantime, the price of sugar had skyrocketed on account of a slave rebellion on the largest sugar-producing island in the Americas, the French colony Saint-Domingue (today Haiti). Klaproth thus stated in his report that "even if calculated against the usually lower prices of foreign sugar in times of peace, the preparation of sugar from beetroot can be done with a significant advantage for the state."[25] At the end of January 1799, both the General Directory and the Academy's directors gave Achard permission to proceed with his sugar production experiments. They established a commission, with Gerhard as its director and also including Klaproth, for the supervision of the experiments.

After a series of further experiments conducted together with Klaproth, on October 31, 1799 Achard submitted a proposal to the Academy's directorate to turn the academic laboratory into a *"Zuckerfabrik"*—a sugar factory. "I wish to use the Academy's laboratory as a sugar factory throughout the winter," he wrote, to produce "a few hundred tons of sugar" from beets. He intended to prove that "nothing stands in the way of the large-scale practicability of domestic sugar fabrication."[26] With that, the last and decisive phase of a large-scale technological experiment began, which would also have profound consequences for the academic laboratory. In the months that followed, the laboratory was completely emptied, retrofitted by plumbers, smiths, and coopers, and equipped with large simmering pans, iron kettles, and further instruments for sugar production. Another part of the laboratory building on the ground floor was used as a room for the crystallization of sugar, and a further room served to store the cutting machines and presses. The beets were stored in the cellar, and a small adjacent building was constructed as well.

In January 1800, the technological experiments in the academic "sugar factory" began, which the king financially supported with about fifteen thousand thalers. Nine workers were hired who were responsible for the handwork, as well as five assistants who supervised the workers and recorded the experimental results. In May, Gerhard, as chair of the commission, reported the first successful results, and at the end of July 1800, the experiments were concluded successfully. Shortly thereafter Achard started to build a commercial sugar factory at a newly acquired estate in Cunern (Silesia). In spring of 1801, Gerhard wrote a summarizing account of the sugar beet project that was positive overall. Friedrich Wilhelm III thanked the commission "for its efforts and zeal in supervising the sugar experiments," which were of utmost importance to national industry.[27]

FIGURE 1.3
Eighteenth-century sugar refinery.
Source: Herzfeld 1994, 177.

EQUIPMENT OF THE LABORATORY

When in April 1801 Klaproth set foot in the laboratory as its new director, he was shocked to find the floor and walls covered in a black-brown sticky sugar mass. And, contrary to Achard's assurances, the cutting and pressing machines, kettles and pans, as well as all the other tools and equipment of the "sugar factory" had not been removed from the building. A year before, Klaproth had sold his apothecary shop in order to devote himself completely to his new academic obligations. Outraged, he wrote to the board of directors that it should immediately see to the task of "swiftly putting the academic house, which is currently in a bad state" back in order, as befitting "the honor of an Academy."[28] The directors decided, in light of the devastation, to set up a new laboratory, which would be housed in a new side wing of the building.

At the time, the laboratory had almost a fifty-year history. In the early eighteenth century, Leibniz had already argued that the Society of the Sciences, the Academy's predecessor institution, needed a laboratory in order to promote the common good, but his plan was approved only decades later, in the reign of Friedrich II.[29] In spring of 1753, the Academy organized the construction of a two-story building to house the laboratory, with residences for the director of the laboratory and the director of the Academy's observatory on the upper floor. The building was located at the Academy's property at Letzte Straße 10 (today Dortheenstraße 10), opposite the Academy's main building, the Marstall. In the spring of 1754, construction of the new building was complete.

The Academy's laboratory largely resembled the chemical laboratory of the University of Altdorf (see figure 1.4).[30] It was visually dominated by fifteen furnaces, including firmly installed large furnaces, smaller portable furnaces, and two chimneys installed at the walls to the right and the left of the entrance door. A large smelting furnace stood in the middle of the room; most of the other furnaces, used for assaying, distillation, and slow "digestion," were placed beneath the chimneys. Various kinds of chemical instruments and mechanical tools—muffles, retorts, coolers, headpieces, crucibles, cupels, a double bellows, tongs, shovels, and so on—were placed upon or

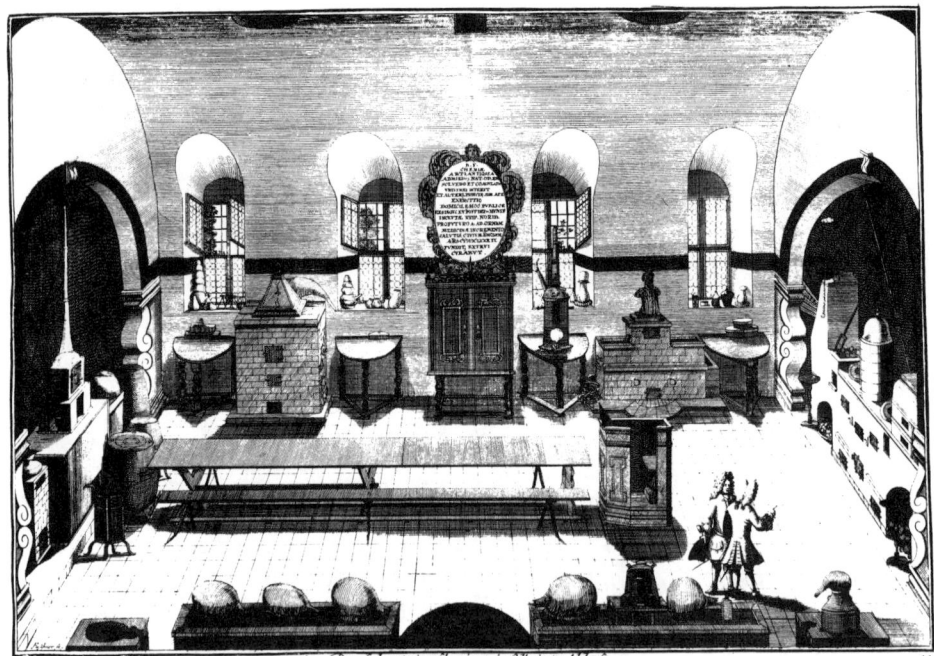

FIGURE 1.4
Laboratory of the University of Altdorf (around 1720).
Source: Niedersächsische Staats- und Universitätsbibliothek Göttingen.

near the furnaces. A table, with small chemical vessels for performing table-top experiments, stood between the laboratory's windows. Further, there was a cistern in one corner of the room as well as a large iron mortar and pestle. In addition to the main laboratory room, there were four smaller rooms, mainly for storing instruments and materials. The bulk of the small instruments and vessels were stored in the building's attic. An inventory listed hundreds of glass and earthenware vessels, including sugar jars, urine bottles, milk bowls, Florentine wine bottles, phials, crucibles, and retorts and receivers of various shapes and sizes. The cellar was used as a coal storeroom.[31]

Andreas Sigismund Marggraf, who became the laboratory's first director in 1753, was responsible for the provisioning of all this equipment, excepting only the large furnaces and chimneys that had been built simultaneously with the construction of the entire building. But he did not equip the laboratory from scratch. Rather, he transferred his own chemical equipment from his pharmaceutical laboratory to the academic building. It took more than ten years until he was reimbursed for this private investment. On the occasion of the laboratory's renovation after the Seven Years' War, he wrote to the Academy's budget commission: "As can be seen from the Academy's archive, everything that exists in the laboratory, a few things excepted, was brought there by me." Knowing that the commission would want to negotiate with him, he immediately added: "It would be fitting for an Academy to possess a complete laboratory equipped with all [necessary] instruments and vessels as well as preparations; this is all the more important as in case of my death my heirs may request [the things] that belong them." In reply, the commission asked him to put together a list of all the things he had "given to the Academy to be used in the laboratory."[32]

In February 1767, Marggraf's inventory was ready. It was an impressive list of instruments, vessels, and materials, which included eight portable furnaces with accessories; larger and smaller coupels and accessories for metallurgical furnace work; more than 400 retorts of different size and materials; two large distillation receivers of green glass and 144 smaller distillation receivers; phials and flasks of white and green glass; numerous very small flasks and retorts; several pots, jars, and bottles with glass stoppers; different kinds of mortars; small and large hammers and anvils; a "machine of glass" with accessories for sun distillation; a "machine" for amalgamation; two tons of white clay, different "preparations and materials," as well as tables and chairs.[33]

Marggraf calculated a total of 322 Reichsthalers for his reimbursement, a sum that may appear modest compared to the annual expenditure for the Academy's observatory, which amounted to 1,295 Reichsthalers in 1768.[34]

However, almost all of the chemical equipment of the time was purchased from ordinary craftsmen and artisans. Marggraf's most expensive regular expenditure after 1766 was not for instruments but for chemical substances and wages. Whereas in the years before 1766 he had only occasionally spent a couple of Reichsthalers for "service," in 1768 his expenditure for wages rose to ca. 85 Reichstalers, that is, more than 40% of his total expenditure of 204 Reichstalers in that year, and it stayed on this level in subsequent years.[35] As for the substances needed for his experiments, he received them from an apothecary named Fabricius and from 1769 also from the well-known Berlin apothecary and chemist Valentin Rose. While Fabricius delivered a broad range of materials, Rose supplied the laboratory with specific chemical substances such as spirit of wine and bitter earth.[36]

Marggraf's equipping of the academic laboratory with his own chemical apparatus is certainly a lucky historical case, providing an exceptional insight into the shared material culture of eighteenth-century pharmaceutical and academic-chemical laboratories. But it is not a unique case. When in December 1802 the Academy's newly built laboratory was ready for Klaproth to move in, he was in exactly in the same situation as Marggraf had been a half-century before. "I have to bring to your attention," he wrote to the board of directors, "that I did not find a single apparatus necessary for an academic laboratory, neither an instrument nor a vessel, nor any chemical preparation; there was no piece of inventory at all, a damaged bellows excepted." No one knew where Achard had stored the laboratory's chemical equipment before he set up his sugar factory. Klaproth suggested that his "own equipment would be sufficient at first," presuming he was reimbursed for it.[37] Thus he equipped the new laboratory with instruments, vessels, and materials from his own pharmaceutical laboratory. The new laboratory wing also contained an auditorium that Klaproth used for his teaching, first in the context of the lecture series of the mining administration and then for his university teaching.[38]

The history of the Academy's laboratory—especially its equipment with chemical apparatus from Marggraf's and Klaproth's pharmaceutical laboratories, and its transformation into a "sugar factory" for carrying out large-scale technological experiments—provides clear evidence that the Academy propagated the "real utility" of the sciences and the "common good," not only in word but in deed. When in November 1807 Alexander von Humboldt helped to reorganize the Academy, he referred once again to the close intertwinement "of academic and commercial-industrial life, [or] of theory and practice."[39] With that he expressed a widespread view of the societal benefits of scientific discoveries and inventions.

2 DISCOVERY AND INVENTION: KLAPROTH

Like Achard, Martin Heinrich Klaproth (1743–1817) exemplifies the figure of scientific-technological expert. Klaproth was an apothecary, a consultant of Minister von Heinitz, a teacher of useful science, and Germany's most renowned chemist in the late eighteenth century. His contemporaries described him as a chemist whose "unrivaled exactitude" and "astute invention of more precise and complete separation methods" enabled him to make numerous new discoveries.[1] In 1789, Klaproth discovered uranium and zirconia (*Zirkonerde*, or zirconium dioxide), and in the years that followed the "earths" (or metal oxides, as they were later known) of strontium, chromium, and cerium. He also contributed to the discovery of tellurium and the earths of beryllium and titanium. Into the last years of his life, he conducted innumerable quantitative chemical analyses and determined the chemical composition of over two hundred substances, most of which were minerals. The analysis in each individual case comprised a veritable research program. The majority of his research took place in his own apothecary's laboratory, as he practiced the pharmaceutical trade until 1800, when he became director of the laboratory of the Academy of Sciences. As we will see, the milieu of the mining industry and Minister von Heinitz's circle of experts also contributed to the advancement of his experiments and discoveries.

At the age of fifteen, Klaproth was apprenticed to an apothecary in Quedlinburg, a small town in the Harz mountains.[2] In 1766, after completion of his apprenticeship, he began his travels as a journeyman, working first in an apothecary shop in Hannover. In 1768, he moved to Berlin and finally to Danzig, where he finished his years as a journeyman. Three years later, he returned to Berlin to take a position in the apothecary shop of Valentin Rose. In 1780, his marriage to a wealthy niece of Marggraf enabled him to buy Marggraf's former apothecary shop. After he passed the pharmaceutical examination with the General Medical Board and received the

FIGURE 2.1
Portrait of Martin Heinrich Klaproth, painting by Carl F. J. H. Schumann.
Source: Deutsches Museum München.

privilege to run an apothecary shop, he immediately began to publish on chemical and pharmaceutical matters.

Being the owner of an apothecary shop with a great reputation opened the door to Berlin's scientific circles. In 1788, Klaproth was inducted into the Royal Prussian Academy of Sciences—although he had never studied at university and had a humble artisanal background (his father was a tailor). Around this time, his public lectures had unleashed a veritable chemistry cult in Berlin. One observer described his chemical lectures, which he gave in his pharmaceutical laboratory, as follows: "You will hardly believe how much the study of chemistry is appreciated here now.... Lectures on chemistry are attended by people from all social classes; what's more, since this winter there are also distinguished members of the fairer sex in the audience." These ladies, he continued, "forsake their coffee and game tables, their assemblies and picnics, to staunchly endure cold and heat, fumes and charcoal dust, and all other discomforts of a chemical workshop."[3] In addition, Klaproth taught at the Medical-Surgical College, in the framework of the mining administration's lecture series, and at Tempelhoff's artillery school, where in 1787 he received the title of professor. In 1810, he was appointed the first professor of chemistry at the newly founded University of Berlin.[4] From this time on, the laboratory of the Royal Prussian Academy of Sciences, whose director he was from 1800, also served as a place for his university lectures. As a university professor, Klaproth continued to do empirical chemistry, offering lectures on experimental chemistry and chemical analysis.

As the proprietor of a thriving apothecary shop, Klaproth produced and sold not only drugs but also chemicals. But he filled even more roles. In 1782, he became an official in the highest Prussian medical authority, the General Medical Board (Obercollegium medico), where he ascended the career ladder to the high rank of a *Geheimer Obermedizinalrat*. By the end of 1786, he joined Minister von Heinitz's circle and became his most important adviser on chemical matters. It was at this same time that his experimental research took a new direction. He began to concentrate on the analysis of minerals, henceforth describing himself as a "chemical mineralogist." Obviously, this reorientation of his research coincided with his entry into the sphere of mining and the world of minerals. Among Klaproth's activities as an expert were his membership in the commission for the evaluation of Franz Carl Achard's beet sugar project as well as his years-long membership in a commission led by Minister von Heinitz for the inspection of the laboratory of the Royal Prussian Porcelain Manufactory (Königliche Porzellan-Manufaktur, or KPM).[5]

One of Klaproth's most important inventions was "uranium yellow," a pigment for the decoration of porcelain. Klaproth took the first step toward this invention immediately after his discovery of uranium in 1789. Further steps followed 1791 and 1792, in collaboration with Friedrich Bergling, a laboratory worker at the KPM. The next section details Klaproth's work of discovery of uranium and invention of uranium yellow, which he performed in his pharmaceutical laboratory. Having a laboratory at one's disposal was just one prerequisite for this endeavor. Another was access to the extremely rare raw mineral pitchblende, from which he isolated uranium. It was the milieu of the mining industry that facilitated the latter. In the summer of 1788, Klaproth traveled to the mining town of Freiberg, where the Mining Academy's famous professor of mineralogy, Abraham Gottlob

Werner, had already attempted to analyze pitchblende. The pitchblende samples for his experiments came from Saxon and Bohemian mines, and he brought them back with him from this trip to Freiberg.[6]

WHAT IS PITCHBLENDE?

In the eighteenth century, pitchblende (uraninite), an extremely rare raw mineral with a rocklike appearance, had only been found in Saxon, Bohemian, and Swedish mines. Before 1789, chemists and mineralogists had only a vague idea of what kind of substance pitchblende might be. The Saxon and Bohemian miners called it *"Pechblende"* because it was black as pitch (*Pech*), while its weight suggested it contained a metal, which in their opinion was only a deception (*Blendung*). By contrast, the chemists and mineralogists who had analyzed pitchblende—exclusively Swedes and Germans who researched in the mining industry milieu—believed that it was indeed a metalliferous ore, and therefore they ordered pitchblende into the class of metals. But which kind of metal pitchblende might contain remained a point of controversy. The experimental effort to isolate the metal was so great, and the amount of metal isolated so slight, that some chemists doubted that they had really achieved its isolation at all.

The Swedish mineralogists Axel Frederic Cronstedt and Johann Gottschalk Wallerius believed, on the basis of their analyses, that pitchblende contained zinc and was therefore a variant of zinc ore. By contrast, Werner thought at first that pitchblende was a variant of iron ore. Later, he believed that it contained the recently discovered earth of tungsten.[7] Klaproth shared the belief that pitchblende contained a metal, but he doubted that the Swedish mineralogists, who had appropriated the German term *Pechblende*, had the same kind of mineral in their hands as the Saxon and Bohemian miners and mineralogists.[8] He detailed the experiments leading to his discovery of uranium in two publications. His publications began with an exact description of the externally observable properties of his samples of pitchblende as well as precise information about their local origin. This was a decisive step toward transregional comparability of minerals. In his publications, Klaproth made it a general rule always to name the origin of mineral samples and describe their observable properties. Hence, he informed his readers that he had experimented with two different varieties of pitchblende from two locations: the first variety came from the Georgwagsfort mine in Johanngeorgenstadt (Saxony); the second was sourced from two mines in Joachimsthal (Bohemia), named Sächsischer Edelleutstollen and Hohe Tannen. Based on their different properties, Klaproth designated the pitchblende from Johanngeorgenstadt as the "black variety with a hard, coal-like appearance" and the samples from Joachimsthal as the "dark gray variety." For the mineralogist, the interesting question was whether the two "varieties" were one and the same "species" of mineral, which had consequences for their naming. For the *chemical* mineralogist, this question could be answered only through determination of their composition by chemical analysis.

CHEMICAL ANALYSIS

A scientific discovery of a new substance, despite its novelty and unpredictability, is no suddenly breaking event. Only through numerous steps of chemical analysis, undertaken over the course of several months, did Klaproth isolate a hitherto unknown component of pitchblende, which he identified as metallic uranium. Klaproth did not have to start from scratch on his discovery. Rather, he stood in a chemical-analytical tradition connected with Prussian chemists such as Caspar Neumann (1683–1737), Johann Heinrich Pott (1692–1777), Andreas Sigismund Marggraf, and Carl Abraham Gerhard. Even so, in the late eighteenth century only a few European chemists carried out quantitative chemical analyses of raw minerals. Furthermore, there were no special textbooks on analytical chemistry, nor was it common practice to detail one's own analytical techniques in such a way that they could be easily replicated, as Klaproth did in his publications.[9]

For his mineral analyses, Klaproth always used several different experimental techniques in sequence. He compared their results and then, much as one would solve a puzzle, he assembled them into a coherent image. First, he decomposed a mineral in the "dry way," trying out several technical variations. Then the "wet way" followed, which involved the dissolution of the mineral in an acid and the subsequent precipitation of a component. Again, this was done with a number of different acids and various kinds of precipitating reagents. The analysis of each individual sample thus required dozens of operations, and each individual operation was often repeated several times to eliminate errors. Therefore, the analysis of a single mineral amounted to a veritable research program.

Klaproth began his dry analysis of the two varieties of pitchblende with the blowpipe test, which had been refined by the Swedish chemists Cronstedt and Tobern Bergman. In the ideal case and with experimental skill, the blowpipe's hot reducing flame drew a fluid metal globule out of the ore. The blowpipe tests yielded lead. But because the amount of lead was so slight, and the properties of pitchblende differed so clearly from those of any known lead ores, Klaproth immediately interpreted the lead as a "foreign admixture."[10] After these preliminary tests, he proceeded to other methods of dry analysis. First, he weighed out a precise quantity of pitchblende, placed it in a retort, and heated it to glowing. He observed the formation of yellow sulfur deposits in the retort's neck as well as a characteristic acrid smell of "volatile sulfuric acid" (sulfur dioxide). In the next experiment, he placed samples of pitchblende on clay plates, covered these with a muffle, and heated the pitchblende to a higher temperature. The smell of volatile sulfuric acid and the loss of weight in the case of both pitchblende varieties showed once again the escape of sulfur and sulfuric acid. Because it was known that metals were often present in their ores in the form of sulfur compounds, this was already a first and important result. But Klaproth did not make any further progress with dry analytical methods.

Next, he went over to wet analysis. He dissolved specific quantities of pitchblende in various mineral acids, using simple glass beakers of the kind available in any pharmaceutical laboratory. In these operations, the rocky matter or "gangue" (*Bergart, Gangart*) would not react, but the characteristic metal or "metal calx" (later termed metal oxide) contained in the ore would dissolve. Like many other chemists and mineralogists of his time, Klaproth appropriated the terms *Bergart* and *Gangart* from miners' language. In his first wet analysis, he ground a sample of pitchblende to a fine powder and poured sulfuric acid on it. The sample hardly changed at all. Muriatic acid also proved to be ineffective. But when Klaproth used the stronger nitric acid, he observed that the sample gradually dissolved. With aqua regia (a mixture of hydrochloric and nitric acid), the strongest mineral acid that could dissolve even gold, the reaction was stronger. The acidic solutions were yellow, and an undissolved part remained, which Klaproth identified as sulfur and gangue. In these experiments, both the steel-gray and the black pitchblende variants behaved in the same way.

Klaproth then turned his attention to the yellow solutions, which he assumed to contain the metallic component of pitchblende. "To learn the nature of the metallic principle [*Grundstoff*] of pitchblende," he wrote, "several experiments were performed with these solutions of nitric acid and aqua regia." His aim was to chemically isolate the metallic component, or "principle," by means of precipitation and subsequently to identify it. As precipitating reagents for dissolved metals and "metal calxes" he used metals and alkaline substances, based on the eighteenth-century affinity tables that provided the relevant basic information.[11]

Klaproth first tried to precipitate the metallic component with zinc, then with iron, and in the third experiment with liver of sulfur (a mixture of potassium sulfides and sulfates). Only in the last-mentioned case he observed a brown-yellow precipitate that settled on the bottom of the vessel. This was an encouraging result, but Klaproth went on to test two additional precipitating reagents: tincture of galls, which Bergman had introduced as a precipitating reagent for iron, and prussiate of potash (*Blutlaugensalz*), also known as "phlogisticated alkali"; in 1781, the Swedish chemist Carl Wilhelm Scheele had used prussiate of potash in his analysis of tungsten ore (scheelite), which led to his discovery of the earth of tungsten (later termed tungsten trioxide).[12] Klaproth's latter experiment was successful. He obtained a small quantity of brown-red precipitate that appeared to differ from any precipitate he knew. He took this observation to be the "surest indication" that pitchblende contained a hitherto unknown metal or an unknown metal calx.[13] In order to obtain precipitates in a larger quantity, he tested further alkaline precipitating agents, and with the "fixed alkalis" (sodium carbonate and potassium carbonate) he finally obtained yellow-colored precipitates in a quantity that was sufficient to carry out further experiments.

THE DISCOVERY OF A NEW SUBSTANCE

The next question was, what was the chemical nature of the yellow precipitate? Like virtually all chemists of his time, Klaproth accepted the phlogiston theory. Based on this theory, he believed that the yellow precipitate was a "metal calx," that is, a metal deprived of its phlogiston.[14] But which metal calx was present in this case? Again, a series of experiments followed to answer this question. Klaproth dissolved samples of the yellow precipitate in various acids, in order to produce salts that could be identified by their particular crystal form. From the kind of salts obtained he could then deduce the particular metal calx they were made up of. His experiments were a success. The yellow precipitate dissolved readily in sulfuric acid, and after the solution was cooled, a "lemon-yellow salt crystallized into small accumulations of columns" developed. Using acetic acid, he obtained clear, topaz-yellow crystals, which "form[ed] regular, quadrilateral, narrow, in part one-half inch long columns, both ends of which led into quadrilateral pyramids."[15] Such salt crystals had never before been observed. The result was clear: a special, hitherto unknown metal calx had been separated from pitchblende.

Klaproth then tried to reduce the yellow metal calx into a metal, not least because the name of a metal calx was derived from the name of the metal it procured.[16] After the blowpipe and all the other traditional methods of metallic reduction had failed, he used a new technique that the Swedish chemists Scheele, Bergman, and Johann Gottlieb Gahn had used fifteen years before in their discovery of manganese.[17] "I grated 120 grains of yellow metallic calx and combined it with flaxseed oil to make a paste," he wrote. He then put the paste "into a pot well-lined with coal dust," and filled "the remaining space in the pot with coal dust, sealed the cover, and put the prepared pot into the porcelain furnace heated to a medium temperature."[18] The description of this final step reads like a culinary recipe, but the idea behind it was that the reduction of the metal calx required the addition of phlogiston; and in analogy to the discovery of manganese, Klaproth used the phlogiston-rich substances flaxseed oil and coal dust to achieve his goal. The porcelain furnace he referred to was a kiln at the KPM, which reached temperatures of up to 1,450 degrees Celsius.

The results of the first experiments Klaproth carried out in this way were ambiguous. Instead of the expected lustrous metal globules, he observed a "loosely joined mass," which had only "some luster." On the other hand, it dissolved in nitric acid, which was a characteristic property of metals. Klaproth then repeated the experiment, this time adding a flux and increasing the temperature. "My expectation was not completely disappointed by the result," wrote the cautious analytical chemist. He had indeed obtained a cohesive metallic regulus, which, worryingly, still had some pores. He then tried to resynthesize the metallic component of pitchblende out of the last-obtained regulus and sulfur, and he succeeded. "The sulfur recombined with the metal," he declared, and the resulting compound was as black as pitchblende.[19] The supplementation of chemical analysis by resynthesis of the original compound (in our case pitchblende) was an established method to make certain that the chemically separated substance was a natural component of the compound and not an artifact. After this last experiment, on September 24, 1789 Klaproth announced the discovery of a new metal to the Royal Prussian Academy of Sciences. He named the metal at first *Uranit*, and then *Uran* (French: *urane*) in allusion to the planet Uranus, which Friedrich Wilhelm Herschel had discovered eight years earlier.[20]

Until 1789, there were altogether seventeen different known metals. The original fixed number of seven metals, which corresponded to the classical seven planets (including the sun and moon), had long since been abandoned. Thus, the discovery of the eighteenth metal no longer called any foundational principles into question. Even so, it was a sensation in the scientific community. Klaproth did not play up his discovery, but also published the experimental uncertainties mentioned above. In the following decades, uranium was not much more than a phantom. Its isolation from pitchblende in sufficient amounts was an extremely complicated process, and thus its availability was long in coming. Only in 1841 did Eugène-Melchior Péligot, a French chemist and assayer at the Paris mint, prove that Klaproth's uranium was in fact uranium dioxide (UO_2). Péligot was the first chemist to isolate true metallic uranium, using completely different methods than Klaproth.[21]

"URANIUM YELLOW" FOR THE KPM

At the time of his discovery of uranium, Klaproth was a member of Minister von Heinitz's inspection commission at the KPM. He thus knew that the pigments used in the decoration of porcelain were "metal calxes." As uranium calx possessed a notably pretty yellow color, the pursuit of its use in the KPM was his next endeavor. Right after his discovery of uranium, he thus carried out experiments to answer the question of whether the yellow uranium calx could be used in the decoration of porcelain and of glass, since glass and porcelain were chemically similar materials. His first experiments were promising. With glass, he obtained colors that varied, depending on the use of further chemical admixtures; one was a brown-gray glass that looked like smoky topaz, while others resembled light apple-green chrysoprase and green emerald. Applied to porcelain, he obtained a "saturated orange color."[22] To complete his invention, he assigned the KPM laboratory worker Friedrich Bergling the task of repeating and perfecting his experiments. As we will see in chapter 4, in autumn of 1792 Bergling could finally report his success in obtaining a beautiful and durable yellow porcelain color. The KPM used the radioactive material, designated as "uranium yellow," for decades.[23]

THE EMERGENCE OF CHEMICAL MINERALOGY

Klaproth's publication of his discovery of uranium also contained new information on the chemical composition of pitchblende, which for mineralogists was as important as the discovery of a new chemical substance. The chemical reform of mineralogy—especially its nomenclature—had even formed the point of departure for Klaproth's analysis of pitchblende. Like many mineralogists with a background in chemistry, Klaproth thought the time had come to replace the traditional, locally varying names for minerals with chemical names that denoted the minerals' chemical composition and would be valid for all mineralogists, thereby facilitating communication. As Klaproth wrote in the opening sentence of his German publication, pitchblende needed to receive a chemically determined name and "a commensurate place in the [taxonomic] system."[24]

In his analysis of pitchblende, Klaproth had first isolated sulfur and then uranium calx. Hence, he assumed that uranium, or uranium calx, was present in pitchblende in the form of a sulfur compound.[25] As sulfur and uranium calx could be isolated from all of his pitchblende samples, he further concluded that the samples were "varieties" of the same "species" of mineral. On this basis, he replaced the old miners' term, pitchblende, with the new chemical name *uranium sulphuratum*, or, in French, *urane sulfureux* (sulfurous uranium). The new term denoted the characteristic component that distinguished pitchblende from other minerals. Hence, Klaproth pointed out that sulfurous uranium was a new "mineralogical species," for which a place in the chemical-mineralogical classification system still had to be found.[26]

Klaproth's transition from "pitchblende" to "sulfurous uranium" is anything but trivial. On the contrary, it exemplifies an ontological shift that had implications for the type of objects investigated in mineralogy and thus for the discipline of mineralogy as a whole. Klaproth's work provides unique insight into the transformation of traditional mineralogy, which was part of natural history, into the discipline of chemical mineralogy, which relied on experimental intervention and the experimental production of chemical-mineralogical species. Whereas pitchblende was a naturally occurring, mineable mineral, which contained other components aside from sulfurous uranium—not least the gangue, which was of technical relevance in smelting works—sulfurous uranium was only one of several components of pitchblende, which owed its material existence (in isolated form) to chemical-analytical techniques. From a chemical point of view, pitchblende was a mixture of different substances, whereas sulfurous uranium was a chemical compound and thus a pure chemical substance.

In the decades that followed, chemical mineralogists replaced the old names for minerals with chemically defined names and chemical formulas. As a result, raw minerals were removed from the paradigmatic core area of nineteenth-century chemical mineralogy. However, they remained important objects of inquiry in the framework of technical chemistry and mining science. For the nineteenth-century mining scientists and technical chemists, the gangue and other constituents of naturally occurring minerals continued to be a disciplinary challenge.

3 PHARMACY AND CHEMISTRY

The eighteenth-century Prussian ministers recruited their experts not only from the Academy of Sciences but also from the artisanal milieu, most notably from the apothecary trade. Klaproth stood at the tip of an iceberg whose invisible part comprised the numerous apothecaries who equipped their own laboratories, manufactured chemical remedies, tested variations and improvements of existing recipes, and performed experiments to invent new chemical remedies. Some of these men carried their investigations further into chemistry and mineralogy and were eventually recognized as chemists. Others, such as Willdenow, developed a stronger interest in medicinal plants, built their own botanical gardens, and eventually became botanists. In late eighteenth-century Germany (and France), around half of the men recognized as chemists were also apothecaries.[1] Hence, the fact that Klaproth's pharmaceutical laboratory was also a site of chemical discoveries and public chemical lectures was by no means an exception.

The apothecary trade had a long tradition in Europe. The first independent apothecary shops were established in the late Middle Ages. In the eighteenth century, apothecaries manufactured and traded in numerous kinds of drugs, ranging from vegetable, animal, and mineral raw materials to ingenious preparations created from dozens of different ingredients. Among the preparations were hundreds of remedies manufactured by means of chemical techniques such as distillation, sublimation, extraction with solvents, and dissolution and subsequent precipitation. Distilled acids, volatile spirits, sublimated essences, extracted vegetable oils, and precipitated salts were substances produced in the pharmaceutical laboratory to be subsequently sold as so-called chemical remedies. These chemical remedies proliferated during the seventeenth century to become a normal part of the stock of drugs in the eighteenth century. As a consequence, the apothecary's craft was in a state of persistent challenge and innovation. There was hardly a single recipe for the manufacture of a chemical remedy that was not questioned, varied, improved upon, or replaced by a new one. And there was hardly a single chemical remedy that was not on the testing bench as a possible adulteration or as yet unidentified material.

The medical and pharmaceutical edicts and pharmacopoeias of late seventeenth- and eighteenth-century Germany took pharmaceutical laboratories for granted (see figures 3.1 and 3.2). Medical edicts, such as the Brandenburg medical edict of 1693, prohibited the purchase of chemical remedies from "vagrants and preparers of chemical remedies [*Laboranten*]" and allowed only those prepared and sold by apothecaries in their own "laboratories."[2] By the early eighteenth century, it was quite common for apothecaries to establish laboratories. Today, pharmaceutical museums exhibit a large range of chemical instruments and vessels used in eighteenth-century pharmaceutical laboratories. The pharmaceutical museum in Heidelberg, for example, exhibits a collection of instruments that were in normal use in apothecaries' laboratories from the seventeenth to the nineteenth centuries. In addition to such collections, written descriptions, inventories, and

FIGURE 3.1
Big laboratory of the Royal Court Pharmacy in Berlin.
Source: Hörmann 1898, 220.

FIGURE 3.2
Small laboratory of the Royal Court Pharmacy in Berlin.
Source: Hörmann 1898, 224.

drawings of pharmaceutical laboratories also demonstrate a strong correspondence between apothecaries' laboratories and chemical laboratories
at academic sites. Above, we have seen that Marggraf and Klaproth transferred their chemical equipment from the pharmaceutical laboratory to the
laboratory of the Royal Prussian Academy of Sciences. This fact attests to
the shared material culture of the pharmaceutical and academic-chemical
laboratory in the eighteenth century. Almost all of the instruments and
vessels used in eighteenth-century pharmaceutical laboratories are also
familiar from chemical laboratories established at universities or scientific
academies. The types of furnaces, retorts, receivers, alembics, jars, bottles,
crucibles, and balances that apothecaries used to make chemical remedies
were the same as those used by academic chemists in their experimental
investigations. What is more, the techniques of chemical-pharmaceutical
production overlapped significantly with those involved in chemical

FIGURE 3.3

Eighteenth-century chemical-pharmaceutical instruments. Source: Apothekenmuseum Heidelberg.

experimentation. Dissolutions, distillations, evaporations, precipitations, calcinations, combustion, smelting, and so on were all both experimental techniques and material production methods.

CHALLENGING CHEMICAL REMEDIES

The eighteenth-century apothecaries often experimented in their laboratories in order to improve the production techniques of a chemical remedy or produce a new variety thereof. These experiments provided a stimulus to extend and deepen chemical knowledge. A case in point is the production of ethers, which were widespread chemical remedies in the eighteenth century.[3]

In the late eighteenth century, ethers already had a long history going back to the sixteenth century.[4] The properties of the pure ordinary ether (sulfuric ether), made with sulfuric acid, were first described in a paper by the German chemist Siegmund August Frobenius (?–1741), published in the Royal Society's *Philosophical Transactions* of 1729–1730. Frobenius introduced the name "ether" (or "aether") and described several of its properties. The substance was volatile, easily flammable, possessed a fragrant odor, dissolved oily substances, and had a "wonderful harmony" with gold.[5] Frobenius praised it as "the most noble, efficacious and useful Instrument in all Chymistry and Pharmacy."[6] In his publication, Frobenius did not include the recipe for making pure sulfuric ether, because he wished to protect his commercial interests. But along with his published paper, he deposited another paper at the Royal Society that gave instructions on how to prepare the substance from oil of vitriol and spirit of wine. After Frobenius's death in 1741, Cromwell Mortimer, the acting secretary of the Royal Society, made the recipe public. From that time onward, pure sulfuric ether began to be used on a broader scale as a chemical remedy, and recipes for its preparation were included in many pharmacopoeias.

Mortimer's publication also initiated attempts to vary the preparation of ether, leading to a cascade of publications by both chemists and apothecaries in ensuing decades.[7] Among the myriad variations, apothecaries and chemists also tried to make ether with acids other than oil of vitriol. The ethers produced using nitric acid, hydrochloric acid, or acetic acid had many properties in common with sulfuric ether, but also differed in some respects. This raised the question of whether these ethers were mere varieties of the same chemical substance or truly different kinds of ether. From a pharmaceutical point of view, this was an important issue, since different chemical substances could have different medical effects. The question was no less interesting from a chemical point of view, since it also touched the understanding of the chemical composition of ethers and of the chemical reactions taking place in their preparation.

The new professional journals for chemists and pharmacists appearing from 1778 onward, such as Crell's *Chemische Annalen*, Göttling's *Almanach*, and the *Annales de Chimie* in France, became a popular forum for sharing information and ideas about ethers.[8] A number of German apothecaries reported their observations in these periodicals, among them many who are forgotten today. To the latter group belonged C. F. Voigt, a little-known apothecary from Erfurt. Voigt first communicated his experience to Göttling before submitting letters for publication. Hence, his case demonstrates the transition from oral to written reports.

In his *Almanach* of 1781, Göttling reported on Voigt's observations concerning the preparation of the ether of nitric acid. The preparation of this ether was known to be dangerous, and Voigt had found a way to avoid the explosions frequently occurring in the process. He achieved this by using saltpeter and vitriolic acid instead of free nitric acid, a specific technique for the mixing of the ingredients, and by carrying out the distillation at a moderate temperature.[9] A glass retort with a tubule (an additional neck melted onto the retort) and a sand cupel were the two more specific instruments needed for the new technique. Göttling further detailed Voigt's technique, including information about the quantities of the ingredients.[10] In 1783 Göttling reported on Voigt's observations concerning another kind of ether obtained from spirit of wine and vinegar. Voigt's original goal had been to make vinegar by fermenting the residue of brandy distillation.[11] When he distilled the vinegar in order to make distilled vinegar, a chemical remedy, he was surprised to obtain a liquor that smelled like the ether of vinegar, for which he had already found a quick preparation method in 1781.[12] In the same issue of the *Almanach*, Göttling reported yet another observation by Voigt, this time on the manufacture of an ether obtained from alcohol and the acid of ants (later "formic acid").[13] Again, Voigt had set out to prepare a quite different remedy, namely an "infusion" made from ants and spirit of wine.[14] In this case, the ether had been known before, but Voigt had found a much quicker way to prepare it. After these reports by Göttling, Voigt apparently had no further hesitation about reporting his trials himself, for the 1784 issue of the *Almanach* included a letter in which he described his new technique of manufacturing the ether of vinegar in great detail, along with a couple of other observations.[15]

Voigt's case shows that eighteenth-century apothecaries carried out experimental trials in order to improve chemical techniques of producing remedies. They often shared their knowledge with chemists and sometimes published their experimental results. Johann Friedrich August Göttling (1753–1809), the editor of the *Almanach*, notably extended his experimental inquires to more systematic chemical studies and eventually became recognized as a chemist. Göttling completed his pharmaceutical apprenticeship in the shop of the renowned apothecary and chemist Johann Christian Wiegleb (1732–1800) and then administered the *Hof-Apotheke* (court apothecary shop) in Weimar. During this time, he became more widely known to the learned public through a book on chemical pharmacy, *Einleitung in die pharmaceutische Chemie*.[16] In 1780, he began to edit his *Alamanch*, whose main goal was "to make pharmacists [*Pharmaceutiker*] acquainted with proper chemical knowledge."[17] In 1785, he relinquished his position at the *Hof-Apotheke* in return for a stipend from Duke Carl August of Saxe-Weimar, allowing him to study at the University of Göttingen for two years. By this time, he was generally recognized as a chemist. From September 1787 until February 1788 he traveled to England at the duke's expense. Like the stipend, the journey was part of an arrangement between Johann Wolfgang von Goethe and the duke, who planned to appoint Göttling to a new chair of chemistry and technology to be established at the University of Jena. In 1789 these plans were realized, and Göttling received a salaried extraordinary professorship at Jena University, shortly after being awarded a doctorate at the same university in acknowledgment of his earlier publications.[18]

Göttling began his studies of ethers by exploring techniques for their preparation and unambiguous identification. In 1779, he published a lengthy paper in Crell's brand-new chemical journal. At the time, he was still a practicing apothecary. The paper provided an overview of the different kinds of ethers, and then discussed theories concerning the formation reactions of ethers. The historical introduction and subsequent systematic division of the paper into eighteen experiments emulated a learned approach to the subject. In the second half of the eighteenth century, many chemists assumed that spirit of wine (alcohol) was composed of water and a flammable component, which was more or less pure phlogiston. This assumption was based partly on the observation of the properties of spirit of wine: it burned easily and without leaving a residue, and it could be mixed with water in any proportion. Further, the assumption relied on the chemical theory of principles, which stated that the properties of a substance were caused by its components or "chemical principles." Here the theory of phlogiston was relevant, which postulated that combustibility was caused by the invisible chemical principle phlogiston.[19] Chemists' assumptions about the composition of spirit of wine were also linked to their explanations of its chemical reactions, such as its reaction with sulfuric acid to form ordinary ether. The renowned French chemist Pierre-Joseph Macquer had long argued that ordinary ether was "nothing else than spirit of wine that is deprived, by means of the vitriolic acid, of a part of its water principle."[20] The chemical affinity between water and vitriolic acid, he had pointed out, caused the decomposition of spirit of wine, leaving an oily component that was rich in phlogiston, namely sulfuric ether. This explanation also supported the assumption that spirit of wine contained water.

Göttling first criticized the fact that many chemists unjustly generalized Macquer's explanation, claiming that all kinds of acids would act in the same way as vitriolic acid in the formation reaction of ethers.[21] He then confronted Macquer's theory with an alternative theory proposed by his pharmaceutical master, Johann Christian Wiegleb. According to Wiegleb's theory, the ethers consisted of the oily part of spirit of wine *plus* a part of the acid used to prepare the ether, and therefore were thought to be products of a simultaneous decomposition—withdrawal of water from spirit of wine by means of the acid—and recomposition from the remaining oily component with a certain proportion of the acid. As a consequence, the ethers prepared with different kinds of acids were different kinds of chemical substances. In order to provide evidence for this theory, Wiegleb invoked the different smell of the ethers prepared with different kinds of acids. Göttling embraced Wiegleb's theory and went on to offer additional quantitative evidence for it. He showed that the weight of the ether of wood acid was greater than the weight of the spirit of wine used to make this ether. "Hence, I dare say that those who claim that a naphtha is merely the oily part of spirit of wine," he stated, "are completely refuted by the foregoing experiment."[22]

It is well known in the historiography of chemistry that Antoine-Laurent Lavoisier used this type of argument to refute the theory of phlogiston. Since the metal "calx" (later termed metal oxide) obtained in the "calcination" (later termed oxidation) of a metal had a greater weight than the metal, Lavoisier argued that the metals did not release phlogiston, but rather combined with a component contained in the ambient air. The case of Göttling demonstrates that apothecaries were long familiar with this kind of reasoning.

CHEMICAL–PHARMACEUTICAL EDUCATION AND TRAINING

As we have seen, apothecaries' manufacture of chemical remedies provided numerous challenges and stimuli for the refinement of chemical knowledge. In the course of the eighteenth century, many apothecaries began to argue that pharmaceutical apprenticeship should be extended to include more formal chemical teaching. The traditional pharmaceutical apprenticeship in eighteenth-century Germany normally took five to six years, followed by six to eight years of service as a journeyman. Apprentices were accepted at the age of fourteen or fifteen after schooling, with primary school education considered sufficient if supplemented by courses in Latin. They were trained in the laboratory and the *Offizin* (the room where remedies were mixed and sold) by a master apothecary, either the owner of the apothecary shop or his or her (in the case of widows) administrator, and by the older journeymen. In addition to simple techniques such as cutting, crushing, mixing, preserving, and storing remedies, pharmaceutical apprentices acquired knowledge about thousands of materials, among them raw materials of vegetable, animal, and mineral origins, traditional Galenic compositions, and chemical remedies. They also acquired skills in weighing and measuring. In the laboratory, they became acquainted with chemical instruments and chemical techniques. Learning chemical techniques relied mainly on the observation of masters and journeymen and practical imitation. But pharmaceutical apprentices also occasionally read books on pharmacy, chemistry, and botany and as well as pharmacopoeias, which were written in Latin. Hence, in the eighteenth century pharmaceutical apprenticeship was not just artisanal training; it also exhibited some characteristics of higher education.

Prussia was one of the first states to undertake efforts to promote and regulate the chemical and botanical education of apothecaries. As early as 1725, King Friedrich Wilhelm I enacted a medical edict ordering that all apothecaries who wished to establish themselves in cities had to pass an examination in practical chemistry at the newly founded Medical-Surgical College in Berlin.[23] They had to demonstrate their chemical and pharmaceutical knowledge and skill by performing several practical *processus pharmaceutico-chymicos* in the classes of the College's professor of practical chemistry, followed by an examination by the physicians of the medical board. In addition, pharmaceutical journeymen were encouraged to attend the Medical-Surgical College's chemical and botanical courses. Johann Heinrich Pott, a chemist and member of the Academy of Sciences, offered the first courses on chemical theory related to the preparation of chemical remedies (*chymia rationalis pharmaceutica*) at the Medical-Surgical College. Experimental chemical courses were given at the nearby laboratory of the Royal *Hofapotheke* (Court Apothecary Shop) by the professor of practical chemistry (*chymia practica*) Caspar Neumann, who was also the Royal Court Apothecary and a member of the Academy of Sciences. Further, the Prussian government encouraged pharmaceutical journeymen to attend a university or other institution that offered formal instruction in chemistry. In this way they were allowed to shorten their seven-year service as journeyman. To give an example, Andreas Sigismund Marggraf was first apprenticed in his father's apothecary shop and in 1726 continued his apprenticeship at the Royal *Hofapotheke* in Berlin, administered at the time by Neumann. He also attended lectures on chemistry by Neumann and Pott at the Medical-Surgical College. Once he completed his apprenticeship in 1731, he served two years as journeyman in different apothecary shops. In 1733 he proceeded to the University of Halle, where he attended lectures on medicine and on chemistry by Friedrich Hofmann (1660–1742) and Johann Juncker (1679–1759). A year later he went to the mining town of Freiberg to learn mineralogy, metallurgy, and assaying in the laboratory of the renowned mining councilor Johann Friedrich Henckel (1678–1744).[24] In 1735, after only four years of service as journeyman, he returned to Berlin to become administrator of his father's apothecary shop, after he completed an examination before the local medical board and swore an oath to obey the Prussian medical edict.

In the last decades of the eighteenth century, German apothecaries' chemical education benefited from the establishment of private chemical-pharmaceutical boarding schools. In 1779, the apothecary and chemist Johann Christian Wiegleb established a "boarding school" (*Privat-Institut*) for the systematic teaching of chemistry in his apothecary shop in Langensalza. Wiegleb fielded requests from families in many European countries "to accept their sons as students of pharmacy and chemistry, provide them with deeper insight into natural science [*Naturwissenschaft*], or teach them commercial business."[25] Until 1798, when his health declined, he taught some forty "students," an average of two every year. Like traditional journeymen, the students lived in Wiegleb's house and participated in the chemical-pharmaceutical operations he carried out in his laboratory, but they also got additional, more formal school education. For the purpose of his teaching, Wiegleb wrote the famous chemistry textbook *Handbuch der Allgemeinen Chemie* (1781). Hence learning by doing, or practice, was combined with theory. Among Wiegleb's students were well-known apothecary-chemists such as Hermbstaedt and Göttling. In 1789, Hermbstaedt opened his own chemical boarding school (*chemische Pensionsanstalt*) in Berlin, which addressed "prospective apothecaries or other curious boys who wanted to become chemists."[26] The students were taught "chemistry in its entirety" as well as neighboring disciplines, such as physics (*Naturlehre*), mineralogy, pharmacy, *materia medica*, chemical analysis, assaying, and metallurgical chemistry. Students performed their analyses in "a laboratory of their own, equipped with the necessary instruments and materials."[27]

By the end of the century, a reform movement for improving the education and training of pharmacists gathered momentum, for which Göttling's *Alamanch* became a public forum. Suggestions for reform agreed in two respects: the request for closer regulation of pharmaceutical training, and the establishment of professional schools for the teaching of chemistry and botany. To give a few examples, one apothecary argued in Göttling's *Almanach* that a good apothecary needed to be familiar with botany and with "all chemical-pharmaceutical preparations, both theoretically and practically." In order to guarantee this, he suggested printed curricula issued by the governmental authorities. Another apothecary proposed public schools, and the young Erfurt apothecary and chemist Johann Bartholomäus Trommsdorff (1770–1837) pleaded for "scientific pharmacy," which included the acquisition of knowledge about natural history and "applied chemistry."[28] Shortly thereafter, Trommsdorff began editing a journal of his own, entitled *Journal der Pharmacie für Aerzte, Apotheker und Chemisten* (*Journal of Pharmacy for Physicians, Apothecaries, and Chemists*), whose main goal was to "extend the scientific study of pharmacy."[29] In the journal's inaugural issue, he published a paper on his own methods of instructing pharmaceutical apprentices, which stressed the systematic teaching of chemistry.

Trommsdorff's teaching of chemistry started with "its pharmaceutical part." The apprentice read chemical books and was instructed further by Trommsdorff in the laboratory, reporting his observations in a notebook that was reviewed by Trommsdorff every week. He was then gradually acquainted "with the entire range of chemistry" by reading Göttling's *Almanach* and Crell's *Chemische Annalen* as well as textbooks on chemistry by Gren, Hermbstaedt, Westrumb, and Wiegleb.[30] In the last year of the apprenticeship, the apprentice had to make his own chemical preparations. Beginning in 1795, prospective apothecaries also had the possibility of training in Trommsdorff's newly founded chemical-physical and pharmaceutical boarding school for boys (*Chemisch-physikalische und pharmaceutische Pensionsanstalt für Jünglinge*), located in Erfurt. This school operated until 1828 and taught more than three hundred students, of whom at least thirteen subsequently established chemical factories. In 1823, the Prussian government announced that attending this school was equivalent to university attendance.

CHEMICAL-PHARMACEUTICAL JOURNALS

The establishment of chemical-pharmaceutical schools in the late eighteenth century was a crucial step forward in the development of chemical pharmacy and the transformation of the traditional apothecary trade. Additional incentives for this development came from the newly published chemical-pharmaceutical journals. Above we have seen how Götting encouraged apothecaries to report their observations in his *Almanach*. Göttling started to publish his *Almanach oder Taschenbuch für Scheidekünstler und Apotheker* (*Almanac or Pocketbook for Chemists and Apothecaries*) in 1780 in the form of a small, bound book, its humble size matching its title.[31] The *Almanach* primarily addressed apothecaries, or "pharmacists [*Pharmaceutiker*]," rather than "truly learned chemists [*würkliche gelehrte Scheidekünstler*]," and its main goal was "to acquaint pharmacists with proper chemical knowledge." Chemical knowledge, Göttling explained, was useful for pharmacists to "banish wrong treatment caused by ignorance as well as adulteration, the use of poisonous vessels, and many additional bad habits existing in apothecary shops."[32]

Göttling divided his journal into two main parts, the first of which contained "brief remarks from chemistry" and the second one "extended treatises." Accordingly, he emphasized that he did not expect all contributions to his journal to be "extended treatises," and that he was willing to publish "each single brief remark."[33] As we have seen above, in some cases Göttling published observations communicated to him orally. In other cases he included short letters in his journal as well as other kinds of brief contributions under headings such as "pharmaceutical and chemical remarks" and "mixed observations." In the eighth volume of the *Almanach*, he proudly announced that readership of his journal was growing rapidly. Again, he reminded his readers that "during daily pharmaceutical-chemical work the attentive worker could always discover phenomena which, even though they were economically useless, very often served to illuminate the darkest hypotheses by delivering facts."[34] An analysis of the content of the *Almanach* shows that the bulk of notes and papers at first concerned two issues: ways of unambiguously identifying substances, and improvements to techniques of preparing chemical remedies. In its first issues, it occasionally published papers concerned with theoretical problems as well. However, after the publication of Lavoisier's antiphlogistic theory in the late 1780s, the *Almanach* also became a forum for discussion of the phlogistic and antiphlogistic theories and the new Lavoisierian chemical nomenclature.

Historians have also pointed out the significant role played by Lorenz Crell's *Chemische Annalen*, published from 1778.[35] Like Göttling, Crell stated in the preface to the first issue of this journal that it would report each single observation, his own role being restricted to that of a collector. The journal's goal, he further pointed out, was to propagate "chemistry as a science" as well as to "exert some influence on various aspects of everyday life."[36] Crell very successfully encouraged apothecaries to read the *Chemische Annalen* and to contribute to it. Among the 564 German subscribers to his journal between 1784 and 1789, 260 (46 percent) were apothecaries, and among its German contributors more than 40 percent were apothecaries as well. What is more, practicing apothecaries were the most active contributors to the journal, with some individual contributors publishing up to 68 papers during the five-year period from 1784 to 1789.[37]

In order to get an idea of what share of the entire pharmaceutical community the subscribers to Crell's journal represented, we can take the Prussian apothecaries as a representative group. In 1798, Hermbstaedt put together a list for the Prussian king that reports a total of 337 apothecaries in Prussia.[38] Apothecary shops were distributed over almost 300 towns, with the highest concentration in Berlin, which had twenty-four shops. Based on the further information that in the period 1784–1789 there were a total of 43 Prussian subscribers to the *Chemische Annalen*, we can estimate that 12–13 percent of the Prussian apothecaries read the journal.[39] Apothecaries' interest in chemical journals provides further evidence for the transformation of the late eighteenth-century apothecary trade into scientifically informed pharmacy, the creation of close links between pharmacy and chemistry, and the development of pharmaceutical chemistry.

4 EXPERTS AT THE ROYAL PRUSSIAN PORCELAIN MANUFACTORY

The Royal Prussian Porcelain Manufactory (KPM) in Berlin was an industrial site that had an extraordinary demand for technical experts. In the late 1780s Minister von Heinitz reorganized the KPM to meet these demands. He improved the training and education of experts and transformed the KPM's laboratory into a true chemical laboratory.

The KPM was founded in 1763, at the end of the Seven Years' War.[1] Production of porcelain was a novel art in eighteenth-century Europe. In the sixteenth century, demand for hard-paste porcelain, imported from China, had increased significantly. A century later, china became the great passion of princes and the upper classes, who also supported efforts to imitate it. In 1709, the German apothecary and alchemist Johann Friedrich Böttger (1682–1719) achieved a breakthrough in the efforts to invent the first European hard-paste porcelain, which was almost identical to the imported china. Böttger had long experimented in collaboration with Ehrenfried Walther von Tschirnhaus (1651–1708), mathematician and director of the chemical laboratory of the Saxon elector Friedrich August I. The two men had received further support from the Saxon mining official Gottfried Pabst von Ohain (1656–1729), who organized supply with minerals. In 1710, Friedrich August I founded the first European porcelain manufactory, located in his castle, Albrechtsburg, in Meissen.[2] By the end of the century, the Holy Roman Empire hosted some twenty porcelain manufactories, among which the state manufactories in Meissen, Vienna, and Berlin were the most famous.

If the initial invention of porcelain had taken years of painstaking chemical experimentation, its subsequent commercial production entailed further, almost endless technical challenges. Almost all porcelain manufactories in continental Europe employed technical experts to cope with these challenges. Among these experts, many had come in contact with chemical science or were explicitly recognized as chemists. Chemical knowledge

FIGURE 4.1
The Royal Prussian Porcelain Manufactory, watercolor from 1818 by Eduard
Gaertner. Source: Stiftung Stadtmuseum Berlin.

and techniques helped to solve technical problems involved in porcelain
manufacture and to understand the nature of these problems. Conversely,
explorations of technical problems and work of improvement and invention extended and refined the existing chemical knowledge.

Chemistry played a role in the production of porcelain paste, glazing,
the preparation of pigments for ornamenting porcelain, and to some extent
in the construction of furnaces and choice of fuel as well. In addition,
chemistry helped to establish new values, such as cleanliness and order
in the laboratory, precision of quality control of materials, standardization
of materials, disciplined observation, and the willingness to write experimental reports. After its reorganization in 1786–1787, the laboratory of the
KPM became a site of frequent experimentation. Interconnections between
chemical and technological inquiry ranged from the application of wet
quantitative analysis to explorative experiments studying the use of oxygen in enameling. The newly established way of teaching experts combined
learning by doing with formal instruction through lectures and reading
of books. The KPM thus became a thriving site of combined practical-
technical, technological, and scientific knowledge.

EXPERTS IN A SYSTEM OF DIVISION OF LABOR

In the 1780s the KPM employed an average of 400 workers, among them children and women.[3] Like state-directed mining, the building of naval fleets, drainage of swamps, construction of canals, and so on, the royal porcelain manufactories belonged to the large-scale industry of the time. Such industry was marked by a high degree of division of labor, a complex organization, and the deployment of outstanding expertise. In the KPM's mills, a group of workers crushed, ground, and purified the hard ingredients for porcelain paste (feldspar, quartz), while a second group of specialized workers blunged and purified the raw "porcelain earth" (a mixture of kaolin and quartz). Supervised by the "arcanists" (*Arcanisten*), another group of specialized workers then made the porcelain paste. In the next step, different groups of workers formed dishes and various kinds of figures out of the paste, which were subsequently fired by specialists at a temperature of 900° C (see figure 4.2). Glazing followed, performed by yet another group of workers. Then came a second round of firing at a temperature of ca. 1450° C, again carried out by specialists, followed by polishing. Furthermore, in the KPM's "pigment laboratory" (*Farbenlaboratorium*) the "laboratory workers" (*Laboranten*) prepared pigments, fluxes, and oils for painting on porcelain, which were used by the painters for overglaze ornamenting and subsequent enameling (at a temperature of ca. 800° C; see figure 4.3).

The following data provide an overview of the KPM's system of division of labor.[4] Around 41 percent of the workers, called *ouvriers*, were highly qualified artisans and craftsmen who had been apprenticed to a master of the manufactory for six to seven years and were relatively well paid. To this group belonged the painters and the artistic workers who fashioned the porcelain ware, the masters responsible for tools and machines, and the laboratory workers. Around 36 percent of the workers were less qualified men in charge of firing, polishing, and so on. The remaining 23 percent were workers without much previous training who performed simple but often physically demanding labor in the mill, furnace house, stable, and so on. The manufacturing process was managed and supervised by a corps of officials (*Offizianten*) who numbered more than thirty in the 1780s. This corps consisted of the KPM's two directors, accountants, and other specialists for management and commerce (ca. 60 percent of the officials in the 1780s) as well as artists and technical experts and their assistants (ca. 40 percent).

The most important technical officials were the KPM's two arcanists. The arcanists were the experts who knew the secret (Latin: *arcanum*) of how to make porcelain. As we will see below in more detail, they were

FIGURE 4.2
Manufacture of porcelain. In the foreground of the building on the left, a man is blunging porcelain earth, which is then dried on a wall in the background; firing is shown at left. In the building on the right, raw material is being crushed in the foreground, with porcelain being formed in the background. Source: Milly 1774, plate 1.

knowledgeable in chemistry, one of them being a university-educated man who was explicitly recognized as a chemist. The arcanists were responsible for a kind of quality control that required them to chemically analyze the ingredients for the porcelain paste and the glaze, calculate the proportions for mixing the paste and the glaze, and supervise the groups of workers who actually made the paste and the glaze and fired the porcelain. The arcanists also performed technological experiments for improving all of these items.

A second group of the technical experts were the KPM's "laboratory workers," usually two to three men. They worked in the pigment laboratory, where they prepared pigments, fluxes, and oils that were used by the painters for overglaze ornamenting and subsequent enameling of the porcelain

FIGURE 4.3
Painting on porcelain ware and enameling: grinding of pigments (right),
painting (left), and enameling in a muffle furnace (in the background).
Source: Milly 1774, plate 8.

ware. The laboratory workers knew the second part of the manufactory's
arcanum: the preparation of pigments and enameling. These experts also
carried out experiments for improving pigments and extending the spectrum of colors. When the KPM was reorganized in 1786–1787, they were
promoted to the group of officials.

Generally speaking, the division of labor in the KPM was organized along
three different axes: first, the type of work (hard physical labor, administrative, technical, and artistic work); second, work that required long, short,
or almost no previous training; third, division into workers (*ouvriers*) and
state officials (*Offizianten*). The KPM was hierarchically structured, with the
king and the director, or the board (from 1786), at the top of the hierarchy,

the officials below them, and the *ouvriers* below the officials. Moreover, the officials and highly qualified workers had assistants.

In addition to the employment of arcanists and laboratory workers in the KPM, the Prussian state administration also invited men of science (*Naturforscher*), mostly chemists and mineralogists of the Royal Prussian Academy of Sciences, to inspect porcelain manufacture, carry out work of invention, and teach and supervise the KPM's experts. Thus, in July 1787 the chemist Martin Heinrich Klaproth became a member of an inspection committee, and shortly afterward Alexander von Humboldt was invited to carry out experiments for technical improvements. The mineralogist and mining councilor Dietrich Ludwig Gustav Karsten, who became a member of the Royal Prussian Academy of Sciences in 1803, also helped to evaluate chemical analyses of porcelain earths.

It is perhaps tempting to regard the division between *ouvriers* and the technical officials, who collaborated with external scientists, as a continuation of the ancient distinction between hand and mind. However, this conclusion would be a simplification that obscures other important features. While such a view would help to illuminate social hierarchy and distribution of power, it would be a hindrance to an appropriate historical understanding of the role played by the state and its officials in the evolution of expertise. The high degree of division of labor in the KPM, and in eighteenth-century large-scale industry in general, also meant the employment of new types of specialized artisans. The role played by knowledge in these experts' handiwork resists the traditional mind-hand divide. On the contrary, careful analysis shows that all of the KPM's experts actually combined mind and hand.

The artistic and technical officials—the master painters and arcanists— were knowledgeable specialists who also worked with their hands. The handwork they did was not related to the immediate production of goods, but rather part of the preparatory technical work comprising the creation of plaster models, design, quality tests, chemical analysis of materials, and systematic experimentation aimed at technical improvement. Regarded from the perspective of the overall system of division of labor, these officials participated in the production process, though in a more mediated way than the *ouvriers*. Their preparative technical work and quality control had an important technical function in the stabilization of high-quality porcelain.

Furthermore, the arcanists' expertise was not categorically different from that of the highly qualified technical *ouvriers*, the laboratory workers (*Laboranten*). This can be seen if we scrutinize and compare the arcanists' and laboratory workers' activities and knowledge (see below). It is further evinced by the careers of the laboratory workers who became arcanists and the fact that the laboratory workers were promoted to the status of officials in the course of the KPM's reorganization in 1786–1787. Archival material about the laboratory worker and later arcanist Friedrich Bergling, presented below, will illuminate the role played by these experts in experimentation, the writing of reports, and the making of chemical knowledge.

EARLY PRACTICES

Under the reign of Friedrich II, the KPM was directly subordinated to the king. Its director, Johann Georg Grieninger, managed daily business, but the king reserved important decisions for himself. The KPM's first arcanist, Ernst Heinrich Reichard, was a sculptor, of whom little is known. After his death in 1764, an accountant named Theodor Gotthilf Manitius (?–1796) took his position. Manitius had first been Reichard's assistant and learned the *arcanum* from him personally. An "instruction for Reichard" from 1763 stipulated that he "had to communicate all parts of the secret to making true porcelain" to Manitius.[5] The KPM's second arcanist, Joachim Duwald (1716/7–1791), was a former potter. He was responsible for the furnaces and the firing of porcelain wares. Unsatisfied with the artisanal status of these two arcanists, Friedrich II ordered the creation of a third position for "a good chymist."[6] In so doing, he followed the example of the Royal Porcelain Manufactory of Meissen, which had employed four arcanists since 1731, of whom two were physicians and chemists. In spring of 1766 the physician Wilhelm Kretschmann (?–1774), from the Prussian city of Halle, became the KPM's "chymicus and arcanist."[7] After his death the physician and chemist Dr. Johann Schopp (?–1797) became his successor.

The archive of the KPM provides little information about the knowledge and activities of its early arcanists and laboratory workers—in strong contrast to the period after Heinitz's reorganization in 1786–1787. In particular, records are lacking about the early arcanists' and laboratory workers' experiments and recipes, again in strong contrast to the period after 1786–1787. The lack of written-down recipes and experimental reports is a strong indicator for personal secrecy. We must thus assume that during the entire reign of Frederick II (1740–1786), the KPM's technical experts kept their knowledge a personal secret.[8]

During these years, the KPM received occasional support from the chemists and mineralogists of the Royal Prussian Academy of Sciences. In the late 1770s, Carl Abraham Gerhard studied cobalt ore, smalt, and *bleu royal* for use as pigments in the ornamentation of porcelain ware.[9] Around the same time, another chemist and member of the Academy of Sciences, Franz Carl

Achard, carried out experiments at the KPM on the preparation of *bleu mourant* (dying blue), a light blue color which was produced by the French at the Royal Porcelain Manufactory at Sèvres.[10] On January 15, 1780, the king could pronounce, "the professor of chymistry Franz Karl Achard created several samples of *bleu mourant*, which must now be subjected to further assays in the manufactory."[11] As a consequence, in 1784 the KPM produced *bleu mourant* and immediately applied it to a new royal dinner service.

The KPM's employment of external chemical experts was not a unique case. Chemists also lent support to state porcelain manufactories elsewhere in Europe. The Royal Porcelain Manufactory of Sèvres, for example, pursued a similar policy. Beginning in 1757, the chemist Pierre-Joseph Macquer performed hundreds of experiments on the making of porcelain, partly in collaboration with the apothecary chemist Antoine Baumé, which eventually led to his discovery of hard-paste porcelain at Sèvres in 1768.[12]

REORGANIZATION OF THE KPM

Shortly after Friedrich Wilhelm II ascended the throne, the KPM was reorganized in several respects. In August 1786, the new king nominated a second director, Carl Jacob Christian Kipfel, who had previously been the supervisor of painters and inspector of materials. In April 1787, Minister von Heinitz convinced the king that the KPM's directors needed to be replaced by an administrative board, the Porzellanmanufaktur-Kommission. The board consisted of the former two directors, a mining councilor named Friedrich Philipp Rosenstiel, who worked in Heinitz's Department of Mining and Smelting Works, and Heinitz, who served as the board's president. The reorganization linked the KPM to the General Directory, the general Prussian state administration, thereby enabling Heinitz to pursue a policy of knowledge transmission and innovation that he had long sought to establish in Prussian mining and metal production (see part II).

Heinitz was an adherent of cameralism and of the Enlightenment, who fully endorsed the high value that these movements attributed to knowledge in the state administration. He was also a practical man, long experienced in the technicalities of mining and manufacture. The minister immediately undertook measures to abolish personal secrecy and institute written reports, as well as steps toward the organization of the laboratory workers' and arcanists' technical and scientific education and training. He also attempted to find new ways to combine technical know-how with insights into things and processes stemming from chemistry, mineralogy, practical mathematics, and other useful sciences.

On July 4, 1787, the KPM's board established a committee for the inspection of the pigment laboratory, which consisted of the mining councilor Rosenstiel, the university-educated arcanist Dr. Schopp, and Klaproth. The committee was charged with carrying out "an exact examination of the pigment laboratory of the Royal Manufactory," in order to gather information on "the quantity and quality of raw materials used for the preparation of the various pigments and on the laboratory workers' techniques for preparation of the individual pigments and the fluxes."[13] An important question was whether the laboratory workers "worked with exactitude and cleanliness and according to the good principles of chemical science." Furthermore, the committee had to check the quality of the instruments and furnaces. With respect to the pigments prepared in the laboratory, it had to answer the question of "whether all of them possessed the required *égalité* [evenness or consistency], so that it was certain that they always yielded the same effects when used in painting."[14] In more modern terms, this was a quest for standardization. Furthermore, the order requested that the inspection committee make proposals for improvements and begin "to perform chemical experiments of its own."[15] Thus the overall goal of the inspection was twofold. The first was to gather information about a part of the *arcanum* that had long been the laboratory workers' personal knowledge. This was part of Heinitz's policy to abolish personal secrecy and transform it into property of the KPM.[16] The second goal was technical improvement and invention.

The committee was expected to do its work during the next three months and to finish its report by October.[17] However, things turned out somewhat differently. The arcanist Dr. Schopp was apparently too old, or not willing, to cooperate. When the committee's first report finally appeared in June 1789, Dr. Schopp was no longer mentioned. Instead the name of another man, Friedrich Bergling, turned up. Bergling was soon to become the major experimenter for the committee.

CHEMICAL INSTRUCTION OF LABORATORY WORKERS

Friedrich Bergling (?–1797) was an apothecary living in Berlin. In January 1788, the KPM's board hired him as a laboratory assistant to be further trained as a laboratory worker knowledgeable in chemistry. He was the KPM's first laboratory worker, later becoming an arcanist, to receive some formal chemical training. For this purpose, Heinitz involved Klaproth. As a first step, Heinitz asked Klaproth to examine Bergling's chemical knowledge and send him an evaluation.[18] Klaproth wrote back that Bergling was "familiar with the part of chemistry belonging to the pharmaceutical discipline" but still "backward concerning the properly scientific part of chemistry." He recommended that Bergling be trained in the laboratory as well as attend his own chemical course, combining the latter with "diligent reading of good textbooks." With respect to his practical training, he further recommended keeping a notebook, in which he should "not only report the work itself but also pay attention, in accordance with chemical principles, to the rationale behind procedures."[19]

In spring of 1789, Bergling moved on to a position as a laboratory worker. One of his first tasks was to write a report on the pigment laboratory. In May, he received the order to "apply the chemical knowledge and experience hitherto acquired to the perfection of the preparation of pigments, the invention of new pigments and fluxes, and determination of the most advantageous use of enamel fire." In addition, he had to perform all the preparations done in the laboratory and "clearly explain the reasons of the applied techniques."[20] A few weeks later, Bergling's report was completed, and it was very critical. The water used for preparing the pigments, the young laboratory worker pointed out, was not pure enough, and he had observed additional sources of impurity. Furthermore, materials were not stored appropriately. For a true chemical laboratory purity of materials, clean instruments and vessels, and well-ordered storage of materials were essential. The laboratory also needed new pots, glass vessels, and another furnace. The KPM's board accepted almost all of Bergling's suggestions. A comment in the margins of the report stated, "this shall be done for the sake of order and cleanliness."[21]

IMPROVEMENTS OF THE LABORATORY

In June 1789, a few days after Bergling had completed his report, Rosenstiel presented the report of the inspection committee, which at this point consisted of only himself and Klaproth. It seems that Bergling's report was the basis for, or even the blueprint of, the committee's report. In the accompanying letter Rosenstiel remarked that his and Klaproth's own views "often coincide with Bergling's ideas."[22] This remark, like many similar ones, is indicative of the committee's collaborative style of working.

Like Bergling, the committee noted a deficit of good furnaces, balances, vessels, cabinets for storing materials, and many tools, "whose lack is utterly unpleasant for a clean chemist [reinlicher Chemist]."[23] Moreover, the laboratory workers often used their bare hands instead of ladles to take substances out of vessels; they did not precisely weigh the ingredients for preparing pigments and fluxes; and many of the materials were impure and not well ordered on shelves.

Klaproth and Rosenstiel insisted, in particular, on the need for the "unchanging evenness of quality [*beständig gleich Güte*]" of materials, which was not guaranteed by commercial materials. The two commissioners stated: "in our opinion the preparation of invariably good pigments for the Royal Manufactory is too important to be always dependent on the accidental quality of their ingredients, based merely on what a merchant has in stock."[24] They recommended quality-testing of all materials and the construction of a storeroom to stock tested ingredients for at least a whole year. Klaproth added further information about the best merchants to deliver materials. He also offered his help to get access to pure copper and tin from overseas. For a manufactory that needed pure materials, there was also the possibility of buying them from a chemist or an apothecary. Thus, Klaproth recommended the purchase of pure sal ammoniac, an ingredient for fluxes, from the chemist Friedrich A. C. Gren, professor at the University of Halle, who "sold an excellent sal ammoniac."[25] He further recommended the purchase of pure nitric acid and a substitute (*Magisterium plumbi*) for minium, a poisonous lead oxide used as a flux, from his own shop.

The report also contained a list of recipes for the preparation of pigments and fluxes. The recipes were short, resembling the type of recipes contained in pharmacopoeias. They first gave a list of ingredients, along with information about their quantities, and then some basic information about the techniques and tools to be used. This part of the report went back to Heinitz's goal to transfer craft secrets from individual laboratory workers to the board of the KPM. In addition, the report mentioned an invention by Klaproth involving the use of "*platina*" (platinum) for ornamentation. Klaproth had recently reported this invention at a meeting of the Royal Prussian Academy of Sciences and had included a demonstration of samples of porcelain ornamented with platinum.[26] He had described his technique of preparing and using platinum for the ornamentation of porcelain and recommended that it be used as a substitute for silver, because silver did not sufficiently coat the porcelain and quickly lost its lustre. A comment by Klaproth in the margins of the report stated that *platina* had already been tested at the manufactory.[27]

A number of the suggestions made by Bergling and the committee were soon put into practice. In August 1789, the son of director Grieninger, who was an assistant to the KPM's board, reported the installation of a large distillation retort, improvements to storage devices, and the purchase of many new barrels, glass vessels, and tools.[28]

WORK OF INVENTION

In April 1791, Bergling finished another long report on the preparation of pigments and fluxes.[29] By this time, he had become the most important experimenter for the inspection committee. The style of his report, especially its presentation of recipes, was similar to that of Rosenstiel and Klaproth's earlier report. Bergling presented 40 recipes altogether for different pigments, fluxes, and oils, which were based on numerous experimental trials and repetitions of trials in which he had varied the proportions of ingredients or the techniques. Many of his recipes were considerably longer than Rosenstiel and Klaproth's because they included more detailed practical information as well as explanations of observations. For example, Bergling explained the smell occurring in a certain operation by reconstructing the chemical reaction that took place. In so doing he used chemical language and concepts such as "displacement" of a component of a compound or combination with "phlogiston." Well into the 1790s, most German chemists, including Bergling's teacher Klaproth, adhered to the phlogiston theory. As we have seen above, Klaproth applied this theory to his chemical analyes, and he hesitated to discard a theoretical tool that had long been useful in chemistry. While he had accepted oxygen as a major substance involved in combustion, he also argued in 1789 that Lavoisier's rejection of phlogiston relied "on weak reasons."[30]

In September 1791, Rosenstiel sent a second report of the inspection committee to the board of the KPM, based on Bergling's report from April of the same year. The opening paragraph of this report reads: "Following the order of your Excellency [von Heinitz], I had a meeting with Professor Klaproth and Herr Bergling, and we have studied in fine detail Bergling's essays about the materials used in the laboratory of the Royal Porcelain Manufactory and about the preparation of fluxes for gold, silver, and so on, and of the pigments."[31] This statement demonstrates again that Klaproth and Rosenstiel regarded Bergling as their collaborator. Although Bergling had a lower social status than the professor and the mining official, there is no indication that either of the latter saw his report as anything other than a respectable contribution to chemical knowledge. The committee's report thus corresponded to that of Bergling. Add to this fact that, from 1791 on, Klaproth and Bergling continually exchanged ideas about improvements of existing pigments. Klaproth made comments on Bergling's suggestions, written on the margins of the latter's report or in the committee's own report. Likewise, Bergling commented on Klaproth's suggestions for improvement. All of this took place in a context in which the two men technically had a student-teacher relationship.

Beginning in late 1791, Bergling and Klaproth performed a large number of inventive experiments. It had long been a goal of the directors of the KPM to extend the spectrum of colors used for ornamenting porcelain. Apart from Klaproth's invention of *platina*, mentioned above, one of the earliest achievements in this respect was Achard's invention of *bleu mourant* in 1780. The committee's new report emphasized that Klaproth and Bergling were now able to prepare the pigment "gold purple" without the use of tin; a darker "dark blue" by adding pyrolusite (natural manganese dioxide); a "dark yellow" with "regulus of antimony" instead of raw antimony; a "light brown" with sublimated zinc instead of calamine; and a darker "chestnut brown" by adding pyrolusite. Klaproth also suggested testing entirely new pigments containing scheelite, lapis lazuli, and uranium; the last-mentioned Klaproth had discovered just two years before, in September 1789.[32] As we have seen in chapter 2, immediately after his discovery he had examined the possible use of "uranium calx" for coloring glass and for painting on porcelain.

In the following months, Bergling continued Klaproth's earlier experiments with uranium calx. In March 1792, he reported the first results of the experiments that had been "proposed by professor Klaproth" and also presented samples of the colored porcelain.[33] Two additional reports followed in September and December of the same year.[34] Bergling apparently succeeded in the preparation of a new pigment based on uranium as well as one with scheelite, which both yielded "a nice yellow color." Thus "uranium yellow" (*Urangelb*) is included in a table of porcelain colours from 1838.[35]

QUANTITATIVE CHEMICAL ANALYSIS

On May 26, 1791, the king pronounced a "new *Reglement* for the personnel of the KPM."[36] The order effectively made the arcanists responsible for the two parts of the *arcanum*, that is, the making of the paste along with glazing as well as the preparation of pigments and auxiliary materials for painting. This degree was fortuitous for Bergling because he had started a series of new experiments to acquaint himself with the work of an arcanist as well. He had performed wet quantitative analyses of porcelain earths stemming from different deposits, and of feldspar, quartz, and some other kinds of stones. He did this with Klaproth and another scientist, the mineralogist and mining official Dietrich Ludwig Gustav Karsten. Karsten had first studied the mining sciences at the Freiberg Mining Academy from 1782 to 1786 and then continued his studies for one year at the University of Halle. In 1789, he became a mining assistant in Heinitz's Department of Mining and Smelting Works, began to teach mineralogy in the lecture series of the mining administration, and in 1792 was promoted to mining councilor. In 1803, he became a member of the Royal Prussian Academy of Sciences. Like Klaproth, he was both a technical expert and a man of science.

Porcelain earth (a mixture of kaolin and quartz) was the most important ingredient for the paste, but it was also a problematic material because its composition varied, depending on the local deposits it came from. The KPM got its porcelain earth mainly from deposits near the town of Halle at Brachwitz, Beidersee, Morl, and Sennewitz. Other sources included Silesia (deposits around Ströbel) and the surroundings of the town of Passau, which had previously been its main supplier. The KPM's feldspar stemmed from Silesian deposits (Krumhübel, Lomnitz, Schreibershau).[37] Feldspar coming from different natural deposits, or even from different parts of the same deposit, also presented a problem because its composition was never exactly the same. The KPM's first arcanists had certainly been aware of this problem, since they possessed chemical knowledge, but it may very well be that they had restricted their quality tests to studies of properties relying on sensory impressions. In the late 1780s, however, the KPM's board agreed that knowledge about the chemical composition of the ingredients of porcelain paste was crucial for the success of manufacture. The wet quantitative chemical analysis of minerals—one of the most recent chemical methods—yielded the most exact knowledge in this respect, and Klaproth was one of the most accomplished experts on this method in the world. Thus around 1790 at the very latest, wet quantitative chemical analysis became a significant technique involved in the KPM's quality control of porcelain earths, feldspar, and other minerals.

In January 1791, Bergling finished his first report on the wet chemical analysis of four samples of porcelain earth stemming from the deposits at Morl, Beidersee, Ströbel, and Passau.[38] He had mixed a sample (1 Loth) of each type of earth with a distinct quantity of oil of vitriol and distilled water, cooked the mixture, purified the remaining material after dissolution with hot distilled water, and filtered it several times. He identified the component not dissolved in this operation as "siliceous earth" and further determined its weight. He then isolated the dissolved components of the porcelain earth through precipitation with a solution of potash, which yielded "argillaceous earth" and a small proportion of "calcareous earth." Bergling thus showed that the four samples contained the same components—"siliceous earth" (quartz), "argillaceous earth" (kaolin), and a small quantity of calcareous earth—but in different proportions. A few weeks later, he reported on similar analyses of feldspar, quartz, and a few additional stones.[39]

The case shows that Bergling, the would-be arcanist, was learning by doing. Furthermore, it shows that the use of wet quantitative analysis for quality controls of porcelain earths was not entirely unproblematic, since the new method had apparently to be adapted to the manufactory's materials. Hence Klaproth and Karsten accompanied Bergling's experiments. Karsten first read Bergling's report and then commented on it in written form. In his comment, he doubted that the KPM's porcelain earths actually contained calcareous earth. "This is impossible," he wrote, "since this mixture would not yield porcelain but rather glass, according to all known chemical experience."[40] He then speculated that Bergling's samples of porcelain earths were not sufficiently purified, that his reagents were impure, and that his utensils were not clean. He also wondered whether the quantity of the samples to be analyzed was too small. Although his report was critical, it ended with the conciliatory remark that the materials involved in Bergling's experiments caused "the most difficulties in their analysis" and that even "very good chemists, who are much more experienced, often made multifarious errors in this case."[41] Thus Karsten spelled out clearly that what Bergling was doing here was chemistry and that chemical analysis was not a ready-made technique to be just "applied" to new materials without further experimentation.

The KPM's board immediately forwarded Karsten's report to Bergling, along with the instruction that Bergling should repeat all of the analyses with larger quantities of substances and greater exactitude.[42] On March 3, 1791, Bergling submitted his written response, which was partly a defense and partly an acceptance of Karsten's objections. He had contacted his mentor Klaproth for this purpose and had "talked to him in detail and also showed the earths to him."[43] The direct personal contact between Bergling and Klaproth apparently yielded support for the former. Bergling pointed out that Klaproth "had examined the [separated] calcareous earth in his presence and found that his results were correct and that the earth was pure calcareous earth." But he had yet another, even stronger argument in his favor: "I took the vitriolic acid from professor Klaproth," he stated, "and I assumed it was certain that it is entirely pure, since it was destined for medical use."[44] Likewise, he had taken "the acetic acid from Professor Klaproth, who had confirmed that it was entirely pure."[45] On the other hand, Bergling also conceded that he had erred in some cases and "that Herr Assessor Karsten was right."[46] He ended by stating: "it will always be the greatest encouragement for my service to be supported in my experiments by experienced chemists."[47] Again, we encounter here a case of interaction between a laboratory worker and scientifically trained officials that displays features of collaboration rather than simple subordination of the former under the latter. What is more, the case demonstrates that the application of wet quantitative analysis for the quality control of porcelain earths required additional experiments. These experiments yielded an unexpected result. They showed for the first time that the KPM's porcelain earths contained a small proportion of calcareous earth.

EXPERIMENTS ON "VITAL AIR" WITH ALEXANDER VON HUMBOLDT

In March 1793, a novel type of experiment was carried out at the KPM, which studied the effect of "vital air" (oxygen) on enameling. The experiment was ordered by Minister von Heinitz, who hoped that the experiment would yield new insights into the process of firing during enameling, and thereby lead to ways to save fuel. He also wanted to test the impact of oxygen on enameling and the quality of colors. The experimenters were Bergling and Alexander von Humboldt, who was a Prussian mining official at the time (see chapter 10). In his report from March 1793 Humboldt wrote: "According to the oral order of his Excellency Fr. v. Heinitz to perform an experiment that studies the influence of the vital air, or the formerly so-called dephlogistated gas, on the smelting of pigments during enameling, we have used two mornings last week for preparing the necessary quantity of vital air."[48]

The experiment was performed in the presence of director Grieninger and Klaproth. Bergling and Humboldt first coated two porcelain dishes with green, blue, violet, and red pigments, put the dishes into a muffle, and then started firing. The furnace, which was presumably the small muffle-type furnace installed in the laboratory, was heated not with wood but with coal. After fifteen minutes, the two experimenters began to introduce the oxygen. After half an hour, when a certain quantity of oxygen had been consumed, enameling was completed. The observed effects were not over-whelming: most of the colors did not differ from those obtained by ordinary enameling, with the exception of red, which had no lustre.

"The success of this experiment," Humboldt concluded, is "at least negatively instructing."[49] He then went on to explain the observations. The French chemists Claude-Louis Berthollet and Antoine François Four-croy, he pointed out, had proven that "metal calxes"—which he defined, according to Lavoisier's theory, as compounds consisting of a metal and oxygen—underwent changes of their color in correlation with their satu-ration with oxygen. This may have stimulated Heinitz's assumption that oxygen might also be useful for the enhancement of colors in enameling. Humboldt objected, however, that the pigments did not change chemically in the process of enameling, although they were metal calxes. The reason for this was that enameling was a vitrifaction, which had the effect that the vitrified surfaces prevented any interaction between with the metal calx and the vital air.

Humboldt briefly considered the possibility that oxygen might have another positive effect: the increase of heat during enameling. Heinitz had long been concerned with the KPM's enormous consumption of wood and with its increasing shortage. He had promoted numerous experiments with coal and new types of furnaces in order to find a solution of the problem.[50] Based on Lavoisier's theory of combustion and composition of gases, Humboldt mused that an increase of heat was indeed to be expected when oxygen was added to coal. But he immediately added that there were financial disadvantages. "Would the use of the gas not be more expensive than the savings on wood?" he asked.[51]

Clearly, Humboldt and Bergling's experiment was a failure. Even so, the case shows impressively that the KPM was a site of explorative experimental inquiry. Minister von Heinitz did not feel beaten by the experimental result. Shortly afterward he wrote to the king: "My attempts to continually perfect the forms and ornamentation of porcelain ... imply the use of the newest discoveries of chemistry."[52] Humboldt and Bergling's experiment indeed involved the newest chemistry, but it failed, just as numerous technological experiments would do in the decades and centuries to come.

CONTINUITY OF CHEMICAL TRAINING

In June 1793, after the KPM's elderly arcanist, Manitius, had retired, Bergling was formally promoted to the position of an arcanist. In the spring of 1795 he completed an inventory, in the form of a table, of all pigments and substances contained in the manufactory's laboratory. As Bergling and the inspection committee had recommended, the laboratory now held its pigments in store for a longer period.[53]

In the same year, the KPM started to sell a new product: "sanitary dishes [*Gesundheitsgeschirr*]."[54] The paste for this new kind of porcelain was made with the same ingredients as the KPM's true porcelain (porcelain earth, feldspar, and a bit of clay) but contained a considerably larger proportion of clay than true porcelain. Because it was fired at lower temperatures than true porcelain, it was possible to use hitherto unused space in the furnaces, which implied a more economic use of fuel. Hence, the sanitary dishes were cheaper and attracted new groups of consumers. This led to a considerable increase of the KPM's revenue between 1795 and 1805.

The new paste was a coinvention of Bergling and a newly employed laboratory worker named Johann Georg Roesch (1767–1821), whose career was similar to Bergling's. Roesch had first been an apprentice painter at the KPM and then attended the lectures organized by the Berlin mining administration (see chapter 9). His teachers had been Klaproth, Karsten, and Hermbstaedt.[55] In 1791, he was hired as a laboratory assistant, and in 1793 he was promoted to laboratory worker. In 1795, after his return from travels to porcelain manufactories and potteries in Saxony, Thuringia, and Austria, he was promoted to the position of a vice-arcanist.

In the spring of 1797, the KPM hired yet another laboratory assistant, Georg Friedrich C. Frick (1781–1848), who was the son of an assayer at the Berlin mint. In the years to come, Frick received a formal chemical education and training similar to what Bergling and Roesch had, first in the context of the lectures organized by the mining administration and later at the newly established Bauakademie (see chapter 15). When the arcanist Dr. Schopp died in June 1797, Roesch became his successor. Heinitz's strategy to establish advanced chemical expertise at the KPM was thus successful. However, this did not happen without some drawbacks.

On August 26, 1797, Bergling died unexpectedly. Three months later, Heinitz entrusted the university-educated chemist Jeremias Benjamin Richter (1762–1807), from the city of Königsberg, with the position of second arcanist. Richter, who is known today as a major inventor of chemical stoichiometry, was also a student of the philosopher Immanuel Kant. Yet he tried to turn back the wheel of reforms, invoking the arcanist's right of personal secrecy. When the board offered additional financial compensation for his recipes, he accepted this, without changing his secretive conduct. Richter never communicated his knowledge, and after his death in 1807 his recipes could not be found.[56] There is a deep irony in this story: the student of Kant—the most famous representative of the German Enlightenment elite—claimed personal secrecy, whereas Klaproth, Bergling, Roesch, and Frick, who shared an artisanal background, supported Heinitz's effort to depersonalize the *arcanum* and implement chemical science at the KPM. From a long-term perspective, however, Richter's attitude was a mere episode. In 1877 the KPM eventually founded a true research and teaching school, the Chemisch-Technische Versuchsanstalt.

We can conclude that the laboratory workers and arcanists of the KPM were experts who possessed extraordinary technical know-how as well as chemical knowledge. The technical challenges and complexity of porcelain manufacture created a demand for experts who implemented chemical knowledge and chemical values. As these experts fulfilled crucial technical functions in the system of division of labor, they had the status of state officials.

By the end of the eighteenth century, the KPM was a site of systematic chemical experimentation. Its laboratory was transformed into a true chemical laboratory where chemical values were respected. The laboratory workers reported their experiments in written form, occasionally adding chemical explanations, and they also explored possibilities of improving pigments and of inventing new ones. The KPM's arcanists applied the newest analytical method of chemistry, that is, wet quantitative analysis, in their quality tests of the ingredients of the paste. Moreover, future laboratory workers and arcanists were trained with the help of chemists and mineralogists, and they also attended the lecture series organized by the mining administration and the Bauakademie. This new way of chemical training and education combined learning by doing with more formal scientific instruction.

If we compare technical experts like Bergling with men of science like Klaproth there are similarities and dissimilarities. The two men possessed chemical knowledge and were experimenters employing the most advanced analytical techniques of eighteenth-century chemistry; they were experts on the materials and techniques involved in the manufacture of porcelain; they shared the goal of improving techniques and inventing new, useful materials. And they even had the same social background: both of them had never attended a university but had been apprenticed to a master apothecary. Unlike Bergling, however, Klaproth was also a famous teacher of chemistry, author of numerous chemical papers, a member of the Royal Prussian Academy of Sciences, director of its laboratory (from 1800), and the first professor of chemistry at the newly founded University of Berlin. In short, while Bergling was a technical expert whose chemical knowledge and skills were more narrowly tuned to porcelain manufacture, Klaproth was also recognized as a chemist who was knowledgeable in all fields of the chemical discipline. In the next chapter, we will further illuminate the figure of technical expert and present an overview of its history.

5 THE FIGURE OF TECHNICAL EXPERT

The eighteenth-century German terms for expert—*Sachkundiger* and *Sachverständiger*—meant a person who possessed extraordinary empirical and practical knowledge about a particular technical field that was, as a rule, at the margins of, or entirely outside, the sphere of regular artisanal professions subject to guild regulations.[1] These were often newly established fields, such as porcelain manufacture, the production of chemical remedies, and colonial trade, or they were part of the large-scale industry of the time that had introduced division of labor, as was the case with silver mining and the construction of fortifications. Because in continental Europe the state administration often managed such kinds of complex practical fields, either through regulations or direct intervention, experts often had close relationships to kings, princes, or ministers.[2] The terms *Sachverständige* and *Sachkundige* thus mostly occur in eighteenth-century texts referring to state-organized practices.

The technical experts at the KPM studied in the previous chapter, Bergling, Roesch, and Frick, were exactly such *Sachverständige* and *Sachkundige*. They were practitioners with unusual expertise in a rare practical area, and they fulfilled specific technical functions in a system of division of labor. Moreover, their technical functions had been introduced only in the eighteenth century, along with porcelain manufacture, and thus were not governed by the time-tested knowledge transmitted in the guild system. Like the famous engineers, inventors, and skilled artists of the Renaissance—among them military engineers, architects, shipbuilders, land surveyors, mine assayers, and itinerant project makers—they were inventive men who contributed to the improvement of manufacture. They often had personal relationships with men of science, attended lectures, read books, experimented, wrote experimental reports, and occasionally published their experimental results.

Compared to the traditional craftsman, the technical expert was a new historical figure who triggered the transformation of the knowledge economy in Europe. His practical and social status undermined the ancient ideological bifurcation between scholar and craftsman, knowledge and skill, or mind and hand. In the middle of the eighteenth century, the technical expert was fully recognized as a new type of producer of knowledge, who combined articulated knowledge with bodily skill. To understand the historical innovation represented by the eighteenth-century figure of the technical expert, a brief overview of its long history is instructive.

Already in the late Middle Ages, special socioeconomic niches arose that created a demand for extraordinary practical and empirical knowledge and fostered innovative projects and inventions. Three main niches can be roughly distinguished: early modern "high-tech" industry; the luxury market along with the expanding colonial trade; and the medical-pharmaceutical professions. These three relatively small areas were hothouses for the development of rare technical expertise and inventiveness. They were also sites in which technical, mathematical, and learned natural knowledge were brought together and entered into a long-term fruitful interaction.[3]

Since the late Middle Ages, several large-scale technical enterprises in Europe had been marked by advanced machinery and an elaborate system of division of labor. The division of labor in these industries created a demand for experts in various tasks of planning and management, which also included activities such as experimentation, measurement, and quality tests of materials. To this early modern high-tech sector belonged silver mining and smelting in the old mining regions of Saxony and the Harz mountains; complex construction projects such as the building of cathedrals, warships with heavy gunnery, fortifications, and siege engines; and hydraulic engineering projects such as the drainage of marshland and the construction of canals, harbors, and dikes.

Early modern high-tech enterprises like these were often financed and managed by the state, ranging from princely courts and city councils all the way to the eighteenth-century technical state administrations. But there were also some cases in which clerical commissioners or private investors were the organizers. Clerical commissioners, or cooperating clerical and state commissioners, usually organized the construction of cathedrals. By contrast, the drainage of the Breemster Lake north of Amsterdam between 1608 and 1612, and the construction of Breemster polder, is a case in which wealthy urban merchants raised capital, sold stakes, and organized a board of overseers that managed and supervised work.[4] In early modern Great Britain, the navy excepted, it was also mostly private, inventive entrepreneurs who were at the forefront of high-tech enterprises.[5]

From the fifteenth century onward, the European market was flooded with exotic medicinal herbs, coffee, tea, porcelain, and numerous other colonial wares, the exploration and imitation of which opened up new horizons of knowledge. In addition to courts, the big port cities and marketplaces trading in goods from overseas developed as centers of technical expertise and innovation. Artist engineers, alchemists, apothecaries, distillers, goldsmiths, inventive carpenters, and so on, living in Venice, Florence, Nuremberg, Antwerp, Paris, or London, became makers and providers of luxury goods for wealthy city dwellers and princes. To give some examples, Emma Spary has shown that distillers in eighteenth-century Paris developed extraordinary expertise in producing luxury goods such as liqueurs; some of them also acquired chemical knowledge and became authors of books on distillation.[6] Studies by Harold Cook, Sven Dupré, and Christoph Lüthy on the Renaissance and early modern Netherlands have illuminated networks of merchants, artisans, and artists promoting trade in luxury goods but also possessing advanced knowledge.[7] In the early modern German states, Nuremberg was a similar center and hometown of the goldsmith Jamnitzer, whose artistic work presupposed outstanding talent as well as natural knowledge.[8] In Elizabethan London, colonial trade and navigation created a market for precision instruments and practical mathematics.[9]

Luxury production in early modern cities took place for the most part in small workshops and laboratories. Distillers made their liqueurs in small laboratories; apothecaries prepared their precious chemical remedies and cosmetics in laboratories that often resembled kitchens; Kunckel produced his gold ruby glass for the court of Brandenburg in a small laboratory, located on the Pfaueninsel near Berlin.[10] But some larger sites of luxury production existed as well. To the latter belonged the production of luxury goods in the Palazzo degli Uffizi in Florence; glass production on the Venetian island of Murano; the royal glass manufactory of Saint-Gobain, which produced crystal glass à la façon de Venise; the Gobelins Manufactory in Paris; the silk industry in eighteenth-century Lyon; and the royal porcelain manufactories of Meissen, Berlin, and Sèvres.[11]

The medical professions and the apothecary trade were closely connected with the luxury market. But they formed a special area insofar as they were placed under state regulation and legislation. The early modern medical and pharmaceutical professions gave rise to experts who traded in medicines and offered medical therapies, supplementing the services of university-educated physicians. Among the apothecaries, surgeons, barbers, midwives, distillers, and women trading with herbs there were many individuals who possessed extraordinary medical competence.[12] In addition to knowledge acquired during apprenticeship training with masters, these individuals also took up elements of medical, botanical, and chemical knowledge distributed through books.

For centuries, the technical experts inhabiting these three niches had been dependent on the favor of rulers and patrons. They often traveled from place to place, the early modern courts being among the most important sites where they could make a living. By contrast, in the eighteenth century technical experts found a stable environment in the modern state administrations. They became expert officials and soon received more formal education and training in engineering schools, mining academies, and similar technological schools. As we will see in part IV, by the middle of the nineteenth century the itinerant inventor and technical expert had evolved into the self-confident group of engineers and *Techniker* that was indispensable for Prussia's industrialization.

II THE MODEL: USEFUL SCIENCE AT MINING ACADEMIES

6 SILVER MINING AND THE FREIBERG MINING ACADEMY

Mining was an eighteenth-century industry that had developed a great demand for experts. In the German-speaking countries, the term *Bergbau* (mining) referred to a complex of technologies and industry that encompassed mines; smelting works; metal processing sites; manufactories connected to the mines (*Bergfabriken*) for the production of cobalt, alum, vitriol, arsenic, and sulfur; glass works; and porcelain manufactories. Silver mining, in particular, was a large-scale industry with mines going to ever greater depths. This was possible with the aid of an ingenious mine drainage system consisting of water wheels, the flatrod system (*Stangenkünste*), and water-lifting pumps, embedded in an infrastructure system of artificial ponds, canals, and water pipes. Such a comprehensive technological system was a rarity at the time. The division of labor had a similarly high level of development. There was a whole range of specialized technical professions for surveying, measuring, testing, and other kinds of mining work requiring expertise, practiced outside the guild-controlled craft professions.

Saxony, Prussia's neighbor to the south, was a centuries-old silver mining region whose economy was heavily dependent on mining.[1] From the 1760s, the Prussian state took Saxony's mining administration as a model for organizing its own mining industry and mining expertise. As silver was an important metal for coinage, the silver mining in Saxony was organized by the state. Thus, maintenance and improvement of mining technology were the responsibility of mining officials. In the course of the seventeenth century, Saxony had introduced the "direction principle" (*Direktionsprinzip*) in mining, which accorded the state nearly unrestricted powers of management, including promotion of technical improvements. In legal terms, the direction principle was a further development of the older royal mining prerogative, which reserved the rights of disposition over mineral resources to the sovereign. The sovereign prince could open and operate mines and smelting works, or lease the land to private entrepreneurs or share companies, the so-called *Gewerken* (or *Gewerkschaften*). The *Gewerken* were obligated to turn over a part of their revenue, either the tithe (ten percent) or a freely agreed upon levy (the *Kuxtaxe*), to the state treasury, which assured handsome revenues for the sovereign prince without his having to take entrepreneurial risks himself. In return, the *Gewerken* obtained the exclusive privilege of exploiting the mines as well as a number of other rights. In the seventeenth century, the introduction of the direction principle reduced the tasks of the private *Gewerken* to the function of creditors who, similar to modern investors in a stock corporation, merely had a share in the earned profits. The members of the *Gewerken* bought shares in mines, called *Kuxen*, and for every *Kux* they received a portion of the profits (the *Ausbeute*). In the case of unprofitable mines, they had to pay a contribution (*Zubuße*) to support the mining operations. However, the management of mining, the hiring of personnel, and all the important economic, technical, and administrative decisions lay in the hands of the state and its officials. The same arrangements applied in the case of the smelting works, which in the sixteenth century had passed almost entirely into the possession of the Saxon state.

For the direction of the mines and smelting works, the Saxon state had created hierarchically organized mining authorities. At the top stood a central administration in Dresden, the Finance and Mining Chamber (Kammer- und Berggemach), which mainly carried out fiscal duties. Technical concerns were usually decided upon in the Central Mining Office (Oberbergamt) and Central Smelting Works Office (Oberhüttenamt) located at Freiberg, which had authority over the sixteen local mining offices in the individual mining districts. In 1782, the Finance and Mining Chamber was integrated into the newly founded Privy Finance Committee (Geheime Finanzkollegium), whereby all its directorial duties and decision making in technical questions were officially transferred to the Freiberg offices or to the local offices.

DIVISION OF LABOR: TECHNICAL OFFICIALS

The division of labor in silver mining comprised a variety of professions and a social differentiation comparable to that of the Royal Prussian Porcelain Manufactory discussed in chapter 4. The hewers, haulers, and all others who performed hard physical labor and had completed a relatively short apprenticeship belonged to the simple mine workers. Mining officials performed the intellectually more demanding tasks for which longer training was a prerequisite. In addition, there were the legal tasks in the context of a "mining state," in which the mining authority had jurisdiction in all mining-related affairs.

The hierarchy of officials can be roughly divided into two groups: the lower to middle "technical officials," and the higher mining officials such as mining councilors (*Bergräte*) and mining masters (*Bergmeister*). The technical officials were referred to as officials "*vom Leder*" (literally "of the leather"), which alluded to the leather apron that the hewers wore as protective clothing and thus highlighted these officials' manual work and technical tasks. By contrast, the higher officials were called officials "*von der Feder*" (of the pen), since they were more strongly occupied with administrative work. In 1765, the Freiberg Mining Academy was founded to provide education and professional training for both groups. In the following, we will take a closer look at the practice and knowledge of mining officials, beginning with the group of technical officials.

From its heyday in the sixteenth century, silver mining in Saxony was largely dependent on the work of the mine foremen, surveyors, assayers, and other technical officials. Even if technical know-how was lacking in the higher ranks of mining officials, the lower and middle ranks of officials secured the technical conditions for production. They served in directorial and control functions in the mines and smelting works. Further, they performed their own specific technical tasks. For these purposes, they had to be on site on a daily basis. The group of technical mining officials included the foremen (*Steiger*), mine managers (*Schichtmeister*), smelting masters (*Hüttenmeister*), mine surveyors (*Markscheider*), mine assayers (*Bergprobierer*), smelting works assayers (*Hüttenschreiber*), engineers (*Kunstmeister*), and "jurors" (Latin: *jurati*, German: *Geschworenen*).

As overseers in the mines and smelting works, the foremen, mine managers, and smelting masters were directly involved in ore mining, ore transport, and metal extraction, respectively. Georg Agricola (1494–1555) characterizes the foreman (*Steiger*) as an official "who controls the workmen of the mine" but also must be "skillful in working wood" himself and proficient in timbering shafts, placing posts, as well as in making, extending, and securing underground structures.[2]

In contrast, the mine surveyors (*Markscheider*) were not directly active in production, but had a crucial function in the planning and organization of mine construction and in the determination of the boundaries between the individual mines.[3] They surveyed the mine terrain both above and underground and then drew up true-to-position ground and elevation plans of the mine areas on the basis of their measurements (figure 6.1). On account of their measurement and drafting abilities, the mine surveyors

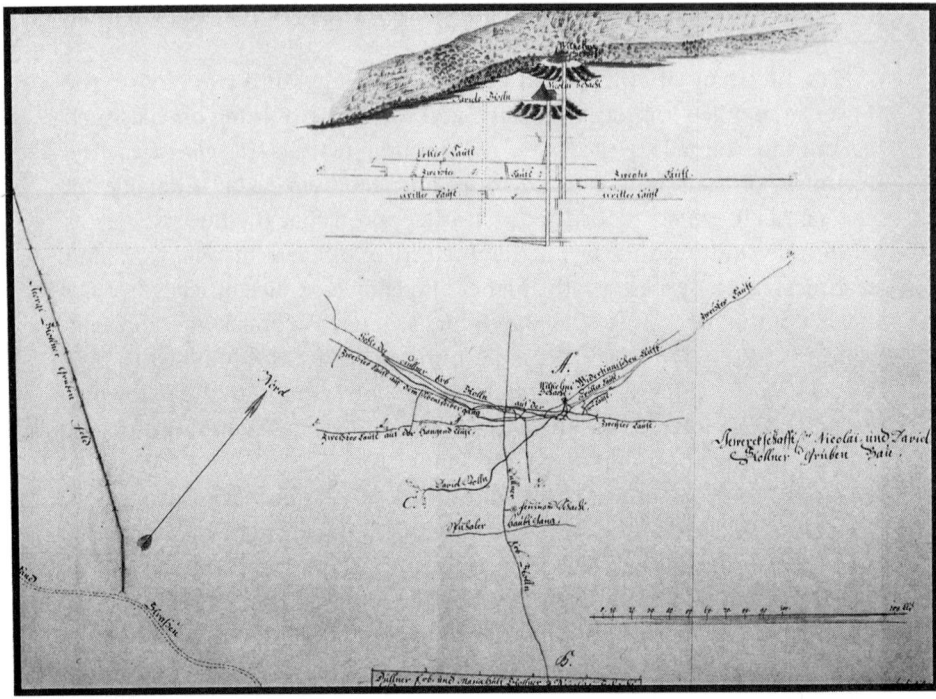

FIGURE 6.1
Eighteenth-century mine plan (plan and elevation), from *Das Goldene Berg-buch* (1764).
Source: Vozár 1983, 117.

were regarded as practical mathematicians. They also possessed outstanding expertise on ore deposits, and could geometrically represent and classify the ore veins according to their position in the rock masses.[4] They received a long and thorough course of training that included instruction in mathematics and was established early on in comparison to the training of the other technical mining officials.

Like the surveyors, the assayers had a key function in the system of division of labor, but were more strongly specialized. Their task was the precise quantitative determination of the metal content of the ores and the metallic melts. To this end, the ores and melts were chemically separated with the help of an assay furnace. Hence, the assayers were occasionally also referred to as *Scheidekünstler* or chemists.[5] In Saxon silver mining, there were several subgroups of assayers: first, the mine assayers (*Bergprobierer* or *Bergwardeine*), who were employed in the local mining offices for testing the metal content of the ores before they were delivered to the smelting works; second, assayers employed by the *Gewerke* who repeated the analysis of the ores; and third, assayers in smelting works (*Hüttenschreiber*), who performed one more assay of the ores. If there were discrepancies between the results of the three analyses, the senior assayer of the Central Mining Office in Freiberg performed a further analysis and made a decision on this basis. The assayers at the smelting works monitored the success of the silver smelting by assaying silver melts at different stages of the process. Like mine survey-ing, assaying was not a direct part of the production processes. Even so, it had an important technical as well as social function within the entire system of division of labor. Socially, it served to balance the interests of the shareholders and the mining authority, as well as the budgetary balancing between the mines and the smelting works.

The *Kunstmeister* organized the construction and maintenance of the min-ing machinery. They were, in today's terminology, mechanical engineers. Their duties also included the construction of instructional scale models of machinery. In Saxon mining, the first such position was created in 1770, at the instigation of the General Mining Commissioner Friedrich Anton von Heinitz, who would later become the Prussian minister responsible for min-ing and the KPM (see above); at the time, it was still common for competent technical experts and administrators to move from state to state. Johann Friedrich Mende (1743–1798), one of the first scholarship students at the Freiberg Mining Academy, was the first to hold the position of *Kunstmeister* and was given the title of "machinery director" (*Maschinendirektor*). Within two years, Mende initiated a surge of innovation in Saxon mining machinery.

The jurors are a borderline case. On the one hand, they spent a substan-tial amount of time in the office and arbitrated, together with the mining master, in legal disputes. On the other hand, they went into the mines to carry out inspections, and thus also had the opportunity to acquire tech-nical knowledge on site. Agricola described them in the context of the sixteenth century as "men experienced in mining matters and of good repute," who "inspect and consider all details, and deliberate and consult with the mine foreman on matters relating to the underground workings, machinery, timbering, and everything else."[6] We will therefore count them as belonging to the group of technical mining officials.

For many centuries, the training of technical officials had been con-ducted within families, with the sons apprenticed to their fathers, uncles, and so on. The mine surveyors are a special case in that their apprenticeship was only possible in larger mining towns. Because the training for surveyors was lengthy and expensive, the Central Mining Office in Freiberg set up a

FIGURE 6.2
Eighteenth-century drawing of an assay furnace, from *Das Goldene Bergbuch* (1764).
Source: Vozár 1983, 132.

scholarship fund in 1702. Henceforth, the state organized the supplementation of practical training by "theory," that is, by formal instruction by means of books, most of which first were manuscripts from which, in the course of the eighteenth century, printed books developed. For the other groups of technical officials, similar arrangements soon followed.[7] Thus the "mining sciences of assaying, mine surveying, silver refining [*Silberbrennen*]," and so on came into existence.[8]

MINING COUNCILORS AND MINING MASTERS

Of the high mining officials, the most important in terms of technical competence were the mining masters (*Bergmeister*), who headed the local mining offices, and the mining councilors (*Bergräte*) and mining commissioners (*Bergkommissionsräte*), who worked in the Central Mining Office in Freiberg. They were subordinate to the *Oberberghauptmann*—the chief mining administrator of Saxon mining. The Central Smelting Office in Freiberg was organized in similar fashion, with the chief smelting administrator (*Oberhüttenverwalter*) at the top of the hierarchy, who was assisted by the chief smelting prefect (*Oberhüttenvorsteher*) and the mining councilors responsible for the smelting operations.

Among these high mining officials, the distribution of manual and intellectual labor corresponded to their social background. Until the end of the eighteenth century, the mining officials at the highest ranks were nobles, and were entrusted primarily with financial matters and paperwork. The chief mining administrator and the highest councilors in the agencies in Dresden were thus always nobles. By contrast, the mining councilors and commissioners working in the Feldberg administrations and the mining masters in charge of the mining offices in the local mining districts were commoners. They were responsible for financial matters, such as the budgeting for the mines, the distribution of yields among the shareholders, and the payment of the *Kuxtaxe* (levy) to the state treasury. Additionally, they frequently entered the mines to carry out technical inspections. In modern parlance, they were financial officials as well as managers and technical experts.

The mining councilors' routine inspections included the monitoring of the technical infrastructure of the mines and the preparatory work for decision making on technical improvements and the financing thereof. Generally speaking, a mining councilor's technical activities varied more both in type and scope than the activities of the technical officials. Whereas the latter were professionally specialized, the high mining officials had a significantly broader field of work to master and also took on numerous ad hoc technical duties. Ideally, a mining councilor possessed both fiscal and technical expertise and endeavored to understand on site the state of operations. The same applies in even greater measure to the mining masters (*Bergmeister*) in the local mining districts, who, alongside financial affairs, were responsible for all the operational management and mining technology in a particular mining district (see chapter 10). After the Seven Years' War, the Saxon authorities undertook measures to live up to these ideals, among which the foundation of a mining academy in Freiberg was the most important.

THE FREIBERG MINING ACADEMY

The Freiberg Mining Academy was founded in 1765 as a school for the education and professional training of Saxon mining officials.[9] It quickly became an international magnet for young men with an interest in the mining sciences. One of the first students was Friedrich Wilhelm von Trebra, the mineralogy and mining technology teacher and friend of Johann Wolfgang von Goethe. In the 1790s Alexander von Humboldt, Leopold von Buch, and Novalis (Friedrich von Hardenberg) studied there.[10] The professor of mineralogy and geognosy Abraham Gottlob Werner (1749–1817) was a celebrity well beyond the borders of Saxony, but the Academy's attractiveness also stemmed from its combination of theory and practice. The theory, that is, classroom teaching, was supplemented by practice in the form of daily work in the mines and smelting works. Hence, an advertisement from 1767 stated that for the teaching of "mining knowledge," the Academy offered both "systematic instruction" in the "auxiliary sciences" as well as "practical training" in mining and smelting.[11] Such a combination of scientific education and practical training was also established at the Austro-Hungarian mining academy in Schemnitz (today Banská Štiavnica, Slovakia).[12]

The course of study in Freiberg was strictly regulated. The morning was reserved for practice, while theory followed in the afternoon. Every student had to write reports on his practical work. These so-called *Elaboratorien* represented the most important divergence of the academic practical training from a conventional artisanal apprenticeship. Otherwise, "practice" was much like an artisanal apprenticeship, consisting of oral instruction, imitation, and physical practicing of manual techniques. In August 1791, the young Alexander von Humboldt gave an idea of his practical work in the mines: "I arrive at 6 o'clock, regularly every day," he wrote to a friend. "I have been doing this work for about three weeks, and at least I am no longer bleeding."[13] The unaccustomed use of pick and hammer in mining for ore had led to bloody hand injuries. A bit later, Humboldt described more fully his activities as an "apprentice hewer" and the "theory" that followed. "I have learned all the menial work with the rock myself, tunneled through my 'hewer's-apprentice layer,' as we say here, and this morning I was still occupied with boring and blasting," he wrote. "Around 11 or 12, I emerge from the mine and now almost all hours of the afternoon are taken up with classes—oryctognosy and geognosy with Werner, mine surveying, assays on silver, technical drawing lessons for plans and machines."[14]

In their practical course, the students got a feel for each operation and why it was being done. They became familiar with the mining techniques, ranging from hewing and ore transportation to the construction of the mineshafts and the system for water lifting. In addition, they had the opportunity to collect minerals and to observe the ore veins and stratification of rocks. Clearly, in the mines natural things and artifacts were intimately connected.

The instruction in the afternoon comprised natural sciences, mathematics, technical drawing, and mining technology. The mixed natural and technological instruction in mining science mirrored the interconnectedness of nature and technology in practical mining. The classes were mainly held in the academy building, either in the lecture hall or in the rooms of the mineral and instrument collections. The laboratories of the senior assayer and of the professor for chemistry were additional teaching sites.

In accordance with the Academy's emphasis on both science and technology, the faculty consisted of professors as well as technical officials. The professors taught mineralogy, geognosy (which later became geology), metallurgical chemistry, physics, mathematics, mining arts, mining law, and mining business management (*Geschäftsstil*); occasionally they offered courses in other subjects as well. The technical mining officials taught mine surveying, technical drawing, the use of mining instruments, and assaying techniques.

As we will see in the following chapters, the composition of the faculty, consisting of both professors and technical officials, became a model for the new technological schools founded in Prussia, most importantly the Academy of Civil Engineering and Architecture (Bauakademie) and the Industrial Institute (Gewerbeinstitut).

THE CAMERALIST DISCOURSE: MINING SCIENCE AND MINING SCHOLARS

The founding of the Freiberg Mining Academy was embedded in a long cameralist discourse about the need for such an institution.[15] In the 1740s, the Saxon cameralist and commission councilor (*Kommissionsrat*) Carl Friedrich Zimmermann had published several essays in which he advocated the collective development of a "science of mining" (*Bergwerkswissenschaft*) and the establishment of a "mining academy" (*Bergakademie*) in order to improve Saxon mining. Similar proposals were contained in the writings of the famous cameralist Johann Heinrich Gottlob Justi, who argued that for the education of "competent and skilled subjects" it was necessary that "the mining sciences" be brought into "a state of greater completion."[16]

According to Zimmermann's plan, the mining academy had to perform the dual functions of a learned society and of a teaching institution. The learned society would be made up of salaried scientists (*Naturforscher*) and tasked primarily with the development of a systematic science of mining, which in Zimmermann's view did not yet exist. This aim required the cooperation of men of science from different fields in a long process of intense public discussion and collaborative work on programmatic scientific texts. In addition, the learned society would write assessments of the proposals for improvements and projects. The teaching of future mining officials was to begin in the second developmental phase of the mining academy, only after the scholars had come to an agreement on the mining sciences. Zimmermann's plan also contained several organizational proposals for teaching and research facilities, including the establishment of a library, an instrument collection, a mineral cabinet, and a chemical laboratory.

Because the financing of such an institution could not be arranged immediately, Zimmermann proposed, as a less costly direct measure, the employment of a "mining scholar" (*Bergwerks-Gelehrter*), who would be both a man of science and an expert on mining. His term *Bergwerks-Gelehrter*—which resonates with our term scientific-technological expert—nicely highlights the ambiguous standing of such a person between the world of the learned and that of the practitioners. Zimmermann specified that the mining scholar would hold experimental lectures for the advanced training of the mining officials and organize a mineralogical-geographical survey of Saxony. Furthermore, he would be an adviser of the mining authority.[17]

The Mining Academy (*Bergakademie*) founded in 1765 in Freiberg, on the initiative of the Chief Mining Administrator Friedrich Wilhelm von Oppel (1720–1769) and the General Mining Commissioner Friedrich Anton von Heinitz, only partially corresponded to Zimmermann's plan. It was mainly a new institution for the teaching of the mining science, but never became a learned society as well. Heinitz had entered Saxon state service in 1763. For the cameralist reformers who supported his appointment, he embodied the ideal of the good cameralist: a highly competent man who served the state and fostered the common good by firmly establishing the place of expertise in state administration. After a short time in Saxon state service, Heinitz became convinced that the mining industry, the mining administration, and the education of mining officials were in dire need of reform. He criticized, in particular, the ignorance and unreliability of the chief mining officials working in Dresden. The new type of mining officials he had in mind would be loyal, diligent, and, most importantly, knowledgeable in scientific and technical matters.

In the fall of 1765 Heinitz formulated a plan that prepared the way for the establishment of the Freiberg Mining Academy. He recommended the creation of three new arrangements for the training of officials without which, he wrote, "we cannot expect to find so soon the good subjects we are lacking in the mining authority [*Bergkollegio*] in Dresden or at the Central Mining Office in Freiberg."[18] In addition, a library, a mineralogical collection, and a collection of models were to be set up, because the "bad apparatus of [existing] models, plans, and ore specimens" in the administration was "unacceptable."[19] In order to train technical officials, Heinitz proposed to establish a new "school for geometrical drawing" as well as a "metallurgical-chymical school." He also pointed out that there were six young Saxon men who could immediately begin a one- to two-year course of instruction at the two schools, at the state's expense. The instructors were to be Johann Friedrich Wilhelm von Charpentier (1738–1805) and Christlieb Ehregott Gellert (1713–1795) at the geometrical drawing school and the metallurgical-chemical school, respectively.

In addition, Heinitz proposed to provide a scholarship for "young cavaliers," mainly young nobles, who were intended for careers as high mining officials. The scholarship, he pointed out, would enable an individually designed five-year training and education, beginning with two years of practical training, followed by two years of theoretical studies of mathematics and sciences at a Saxon university; for the fifth year, a trip to foreign mining regions was foreseen. The year abroad, by far the most expensive part of the training, was a particularly important matter for Heinitz and part of his policy of knowledge transfer.[20] All in all, the second part of the plan indicates that Heinitz initially did not intend to found a comprehensive mining academy, let alone an academy designed for the education and training of both high and low mining officials.

Even so, the two schools Heinitz had proposed were soon established as parts of the Freiberg Mining Academy, which combined practice and theory in one place and, further, educated both high and low mining officials. Beginning in 1766, Gellert held lectures on metallurgical chemistry, while

Charpentier taught mathematics, mechanics, as well as geometric plan and profile drawing and perspective drawing. In addition, the assayer Johann Andreas Klotsch taught assaying, the mine surveyor Carl Ernst Richter gave courses on mine surveying, and the inspector Christian Hieronymus Lommer taught the technical parts of mining science, as well as mineralogy in the mineralogical collection. These three instructors were former beneficiaries of the fund set up for the training of technical officials in 1702.[21] In addition, there was also the practical training in the mines and smelting works. The young Trebra, who came to the Freiberg Mining Academy in May 1766, reported later that he found himself among "many people who were quite well versed in the practical side of mining" and that "the lively lessons in the mines," which he "completed with much diligence," were "the best part of all."[22] In the first few decades of the Academy, there were no entrance examinations or criteria for acceptance, nor was there a required curriculum with a fixed length of the course of study.

GEOLOGY IN THE CONTEXT OF MINING: WERNER

In 1775, the Freiberg Mining Academy hired Abraham Gottlob Werner as the new instructor for mineralogy and inspector of the mineralogical collection. Within a few years, Werner became the most prominent lecturer of the Academy, renowned in particular for his teaching of the new field of geognosy, which would soon develop into geology.[23]

Werner, who said of himself that he treated his objects of study "not only as a geognosist, but also as a miner," was certainly no scholar in an ivory tower.[24] The son of an ironworks inspector, he had worked from 1764 to 1769 as a *Hüttenschreiber* in an ironworks. As we have seen above, the *Hüttenschreiber* was responsible for the chemical assaying of the ores on behalf of the smelting works as well as the controlling of the melting process by means of assaying the intermediate products. From 1769 to 1771, Werner studied at the Freiberg Mining Academy, followed by three years studying law at the University of Leipzig. This combination of studies paved the way for a high-ranking position in the Saxon mining administration, but Werner immediately returned to the Mining Academy of Freiberg to teach mineralogy and mining technology. From 1782, he extended his mineralogical lectures to geognosy, including the science of ore deposits.

Werner's geognosy (*Geognosie*) represents an early form of geology, a young science that first developed in Werner's lifetime. Historians of geology agree that mining contributed significantly to the rise of geology.[25] Mining provided outstanding opportunities for systematic geological

FIGURE 6.3
Portrait of Abraham Gottlob Werner.
Source: Hasse 1848.

observation. What is more, geology or geognosy was a central part of the
mining sciences taught at the mining academies. Hence, in addition to
the practice of mining, the teaching of mining science in the second half
of the eighteenth century was the immediate context in which geology
developed. A closer look at this discipline sheds light on the interdependence of the development of the useful sciences and the specialized natural sciences.

Werner's geognosy centered on the identification and classification of the rocks (*Steine*) and the rock masses (*Gebirge*) that make up the earth's crust as well as the study of the structure of rock masses. It also studied the structural regularities of ore deposits, which was linked seamlessly with the art of surveying. Like Werner, many persons came to geognosy from the field of mineralogy. Mineralogy, with a tradition going back to antiquity, investigated hundreds, or even thousands, of kinds of minerals that frequently were present in tiny amounts. In contrast, Werner's "rocks" were macro-objects, and they were few in number. Werner defined them as the objects that make up "the mass of our solid earth."[26]

Werner's "rocks" belonged to a quite particular realm of experience: that of mining. The experiential world of the miners was constitutive for Werner's geognosy in a fundamental way: it stimulated a shift of perspective from the multifarious miniature objects of mineralogy to the large-format objects of geognosy. This is not least evidenced in Werner's terminology. Instead of *Steine* (*rocks*), Werner also spoke of *Gebirge* (rock masses), hence calling his new field the theory of rock masses, or *Gebirgslehre*.[27] In doing so, he adopted an expression from the miners' language, in which the term *Gebirge* signified the large rock masses from which the miners extracted ores. When a miner said he was going into the *Gebirge*, he meant he was entering the subterranean rocky world, independent of whether the mine in question was in fact in the mountains (*Gebirge* in geographical nontechnical usage) or in flat land. In the framework of geognosy, the miner's term *Gebirge* was placed in a new context of concepts and hypotheses about the structure and genesis of the earth's crust, which also changed the meaning of the term.

Mining provided, moreover, concrete visual aids for geognostic descriptions of the structure of rock formations. As Werner observed, the "knowledge of the behavior of the rock strata" was "indispensable" for the geognosist.[28] In the pits, it was possible to directly observe segments of the ore veins and layers of adjacent rocks along with interspersed minerals. The large-scale course of the ore veins and the large-scale extension of rock strata, however, were not directly observable. In Werner's words, the man of geognosy could "not see the interior and all of this construction in its entirety."[29] But here, the mine plans (*Grubenrisse*) were helpful.

Mine plans were plan and elevation diagrams that gave a comprehensive overview of the course of the subterranean veins and rocks along with mined cavities, shafts, tunnels, and the above-ground mine buildings. Drawn by the mine surveyors on the basis of their measurement data, they were true-to-position schematic pictures that were used by the mining authorities in the planning and organization of mining operations, as well as for the drawing of property and usage-rights boundaries. Whereas the first known mine plans from the mid-sixteenth century still bore the unmistakable signature of the individual surveyor and often included small realistic details, such as pictures of mined rock and horses, later plans were more abstract and standardized. These plans aimed to bring the subterranean mine cavities into relation with the earth's surface and on this basis portray them as spatial formations "that follow vertical, horizontal, or spatially inclining rocky structures." There were small-format representations and large-format mine plans with panoramic representations, which could be several meters long. According to the historian of mining Christoph Bartels, the large-format plans were special in that they provided "a record of large-scale relationships and a documentation of their individual structures as parts of functional networks."[30] Thus, the late eighteenth-century concept of rock strata was already prefigured in the mine plans that developed from the sixteenth century.

In addition to the identification and classification of types of rock masses and the representation of their stratification, Werner also provided a theoretical explanation of their genesis. Consequently, his concept of "rock formation" (*Gesteinsformation*) had both a morphological-structural and a

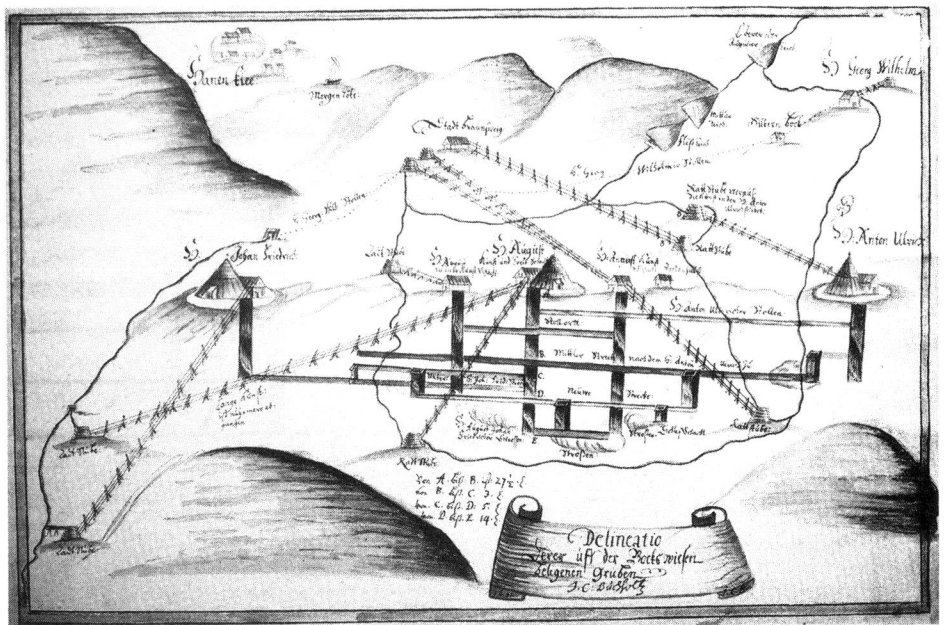

FIGURE 6.4
Mine plan from the Upper Harz (1681).
Source: Bartels et al. 2008, 152.

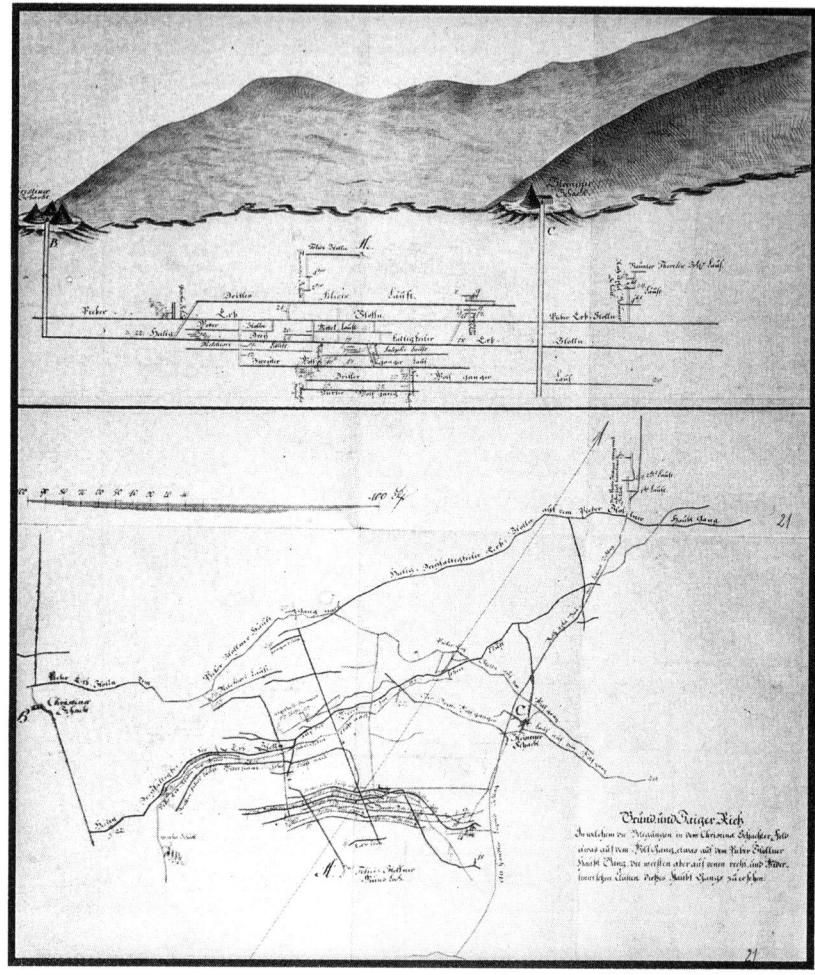

FIGURE 6.5
Eighteenth-century mine plan (plan and elevation), from *Das Goldene Bergbuch* (1764).
Source: Vozár 1983, 59.

historical dimension. In other words, Werner regarded a specific rock formation within the earth's crust not just as a morphological-structural unit but also as a unit in a historical-genetic sense, because it was formed in a specific geohistorical period and in the context of a unitary natural process. Werner aimed to explain this process with his "Neptunist theory."

According to this theory, nearly all rocks were products of mineral sedimentation in the oceans.[31] The exceptions were just a few, most recently formed rocks that were created by volcanic activity. The structure of the earth's crust, consisting of various rock formations, was the result of two great historical phases of sedimentation. In the first phase, when the primordial ocean still covered the whole surface of the earth and was enriched with minerals, a chemical sedimentation (or precipitation) occurred. Typical products of this first phase were crystalline rocks such as granite, gneiss, and schist. Accordingly, these were regarded as the oldest and most primitive types of rock that formed the crystalline basement, which Werner referred to as "primitive rock masses" (*Urgebirge*). Werner dated the first phase of the formation of the basement rock to about a million years in the past, thereby radically diverging from the Biblical history of creation.[32]

Further, Werner's Neptunist theory posited that most other types of rock and rock masses were formed in a second geohistorical phase, in which different climatic conditions predominated that promoted the erosion of the *Urgebirge*. The erosion products slowly formed sediment strata in the oceans, mostly through purely mechanical deposition, and occasionally also through mechanical-chemical processes. The sediment strata were subsequently transformed into the flat-lying stratified rocks *(Flötzgebirge)*, which mostly consisted of sandstone, shale, limestone, and chalk. A special product was the coal seam, in which organic fossils were often deposited.[33] Werner supported this part of his theory in particular with observations of the tectonic structure of the Harz, the Thuringian Forest, and the Saxon Ore Mountains.

USEFULNESS OF GEOGNOSY

Werner regarded geognosy as a part of the science of mining, and this view was broadly shared by the teachers of the Freiberg Mining Academy and by the Saxon mining authority. In his *Neue Theorie von der Entstehung der Gänge, mit Anwendung auf den Bergbau* (*New Theory on the Formation of Veins, with Its Application to the Art of Working Mines*), Werner explicitly addressed the relationship between geognosy and mining, pointing out that his theory of the formation of veins would contribute to "geognosy, the history of our globe, as well as the art of mining." It would provide knowledge of the origin of the veins and their location in the rock formations, which enriched geognosy while at the same time being useful to prospecting for ore deposits.[34] Was this merely the rhetoric of utility, or can Werner's assertions be substantiated?[35] A closer look at his statements shows their high level of reflection and argumentation, thus the opposite of what one commonly would brand as shallow utilitarianism and a naïve faith in progress.

Werner understood the systematic collection of empirical knowledge as a necessary step toward formulating empirical rules, which would be the basis of a theory of geognosy. Thus, in his *New Theory of the Formation of Veins*, he described the composition and spatial orientation of the ore veins in the Freiberg mining region. In addition, he also reported on empirical knowledge of veins in other mining regions. On the basis of his Neptunist theory, he went on to hypothesize that the composition and orientation of the veins were causally related to their historical genesis and were the result of the development of rock formations in the earth's crust. Werner believed that once the nature of the veins and the structural characteristics of the rock formations were better researched by means of detailed empirical study, it would be possible to formulate more refined hypotheses on their historical formation, so that a science of mineral deposits would ultimately result from the interplay of empirical knowledge and theory.[36] He knew very well, however, that all this lay in the future. Geognosy was still in its infancy—Werner was himself one of the founders of the field—and empirical knowledge for a science of mineral deposits and veins first had to be worked out step by step.

The support for geognosy at the Freiberg Mining Academy and in the Saxon mining administration was an expression of a long-term strategy to make scientific knowledge useful for future mining.[37] A brief comparison with the traditional practices for prospecting illuminates the change that was aimed for. The search for new ore deposits was traditionally considered a game of chance. Although there were occasionally empirical indications of a subterranean deposit, such as the presence of certain types of plants and mineral vapors, these were by no means reliable indicators. Only through exploratory digging and excavation, in the late eighteenth century through drilling as well, could one gain more certainty about the presence of ore deposits. The latter was a time-consuming empirical procedure. An alternative consisted of the expansion of existing mines or the reopening of closed ones.

In the eighteenth century, dowsing rods were still occasionally used to find new deposits, although their use had been controversial for centuries. In *De re metallica* (1556) Agricola had already discussed the pros and cons of using dowsing rods to prospect for ore deposits and ultimately advised against their use.[38] Thus it was not the Enlightenment that first questioned the dowsing rod, but rather the miner's own experience, and this since the sixteenth century. In their eyes, the dowsing rod was unreliable because the successes and failures were at best equal. Other motives accompanied skepticism about the dowsing rod as well, such as the assumption that disreputable magic forces were at work. By contrast, the proponents of the dowsing rod argued that an as yet unknown natural force in the ore deposits caused the attraction of the rod, as long as the user held it correctly and was devoid of "inherent peculiarities" that could impede this force. All things considered, for Agricola this led to the conclusion that a good miner who is "prudent and skilled in the natural signs" understands that "a forked stick is of no use to him, for ... there are the natural indications of the veins which he can see for himself without the help of twigs."[39]

Men of science such as Werner, and reform-minded mining officials such as Heinitz, disapproved of the use of the dowsing rod for yet another reason: because the dowsers protected their knowledge as personal secret knowledge.[40] As we have seen in chapter 4 on the Royal Prussian Porcelain Manufactory, Heinitz pursued a systematic strategy to abolish the personal secrecy of technical experts such as arcanists and laboratory workers, whom he obliged to write down their recipes. He devised this years-long strategy already in the context of mining during his time as the General Mining Commissioner in Saxony. The dowsers did not belong to the corps of mining officials, and they were thus not obliged to communicate their knowledge to the mining authorities. Furthermore, there was no organized course of training for dowsers. The secrecy of their knowledge clearly contradicted Heinitz's aim of knowledge transfer within the corps of mining officials. It also contradicted the scientific ethos of the Freiberg Mining Academy. Hence, there was no place for dowsers in the Mining Academy.

WERNER: A SCIENTIFIC-TECHNOLOGICAL EXPERT

In the late eighteenth century, Werner was the celebrity among the professors of the Freiberg Mining Academy. He was one of the founders of geognosy and played a central role in the organization of the science of mining and its teaching at the Academy. In addition to mineralogy and geognosy, he also taught mining technology. Further, he supervised the students' practical work in the mines and smelting works. Cleary, Werner was a man of science. But he was also a mining official and an expert in mining—in German, a *Bergkundiger* or *Bergverständiger*.

In his capacity as a mining official, Werner participated in the practical-technical work of the mining authority. While he was the inspector of the mineralogical collection, his tasks matched his humble rank as a technical official. In 1792 he moved on to the higher rank of a mining commission councilor and a member of the Central Mining Office in Freiberg. From that time, he was responsible for the mine drainage system, including the system of artificial ponds, canals, and water pipes. From 1791 to 1811, the Central Mining Office also commissioned him to direct Saxony's geological survey, the aims of which included mapping as well as the improvement of prospecting for mineral deposits. Werner divided the landscape into ninety-two districts and assigned his students the task of investigating the districts on their own, and preparing cartographic representations and written geological descriptions. The historian of geology Martin Guntau has highlighted the mixed technological-scientific purposes of the survey, stating that Werner "envisaged a systematic representation of the land that would be the basis not only for the documentation of the existence of sought-after deposits, but also for making a mosaic-like synoptic picture of the geological conditions in Saxony."[41]

Werner was a mining official, practical and investigative mining expert, and man of science in one. Like Achard and Klaproth, whose technical activities, scientific inquiry, and work on invention have been detailed in part I, he was a scientific-technological expert. Time and again, the work such hybrid experts did in the framework of the Royal Prussian Academy of Sciences or the Saxon mining industry led to fruitful interactions between scientific research, technical work, and technological problem solving. As we have seen above, the historical actors referred to such hybrid experts of mining as "scholars of mining" (*Bergwerks-Gelehrte*).[42] In the next chapters, we shall encounter similar figures in Prussian mining. Their activities provide additional evidence that geology developed in the milieu of mining, and coevolved with attempts to improve the understanding of the natural conditions of mining.

7 MINING AND MINING EXPERTS IN PRUSSIA: GERHARD

After the Seven Years' War (1756–1763), Prussia introduced the direction principle into its mining sector, following the model of state-organized mining established in Saxony, although iron, not silver, was at the center of the Prussian mining industry. Iron was considered to be crucial for the military, and in the long run it was a key factor in the mechanization of industry. Replacement of wooden machines by machines built of iron would mean the use of mechanisms with higher pressure, more effective power transmission, and unprecedented precision.

As we have seen above, the direction principle entailed that state authorities were responsible for mining equipment and organization. This change in mining laws and policy was the immediate incentive for the establishment of a hierarchically structured mining administration in Prussia, with the Department of Mining and Smelting Works in the General Directory at the top and the local mining administrations below. The local Prussian mining districts were far away from the administrative center in Berlin, reaching from the *Mark* in the west (the region around River Ruhr and the city of Bochum, ca. 450 km from Berlin) to the district around Halle and Magdeburg in the south, the northern parts of the *Neumark* in the province of Brandenburg, and Silesia in the east, some 300 km from Berlin. Hence, inspection tours to the local mining districts were part of the regular technical duties of the high mining officials working in the central mining department in Berlin.

The Department of Mining and Smelting Works in Berlin was headed by ministers who actively promoted reforms, first by Ludwig Philipp Freiherr vom Hagen (1724–1771), then by Jacob Sigismund Waitz von Eschen (1698–1776), and from 1777 by Friedrich Anton von Heinitz. These ministers became role models for knowledgeable, industrious, and loyal officials. They also promoted the teaching of useful sciences. Hence, in 1770 the Department of Mining and Smelting Works organized a lecture series on useful science. In the decades before the establishment of the Mining Department, the General Directory had occasionally asked experts for technical advice and local inspections of mines and smelting works. For example, in 1754 the mineralogist and chemist Johann Gottlob Lehmann, who had carried out several commissions for the Prussian authorities in the preceding two years, received the honorary title of mining councilor (*Bergrat*) and became a member of the Royal Prussian Academy of Sciences in the following month. But not being a salaried official, he soon left Prussia to accept a paid position as professor of chemistry and director of the natural collections at the St. Petersburg Academy.[1] The first mining councilor to make a lifelong career in the Mining Department was the mineralogist, geologist, and chemist Carl Abraham Gerhard (1738–1821). As he climbed the career ladder, younger men succeeded him in his original position, first the Swedish mineralogist and geologist Johann Jacob Ferber in 1786, then the mineralogist and former student of the Freiberg Mining Academy Dietrich Ludwig Gustav Karsten in 1792.

The son of a parson, Gerhard was born in Silesia. In 1755, at the age of seventeen, he moved to Berlin, where he matriculated at the Medical-Surgical College. Among his teachers were the chemist and mineralogist Johann Heinrich Pott, who was also a member of the Academy of Sciences, and the botanist Johann Gottlieb Gleditsch. During his studies in Berlin, Gerhard may also have met Johann Gottlob Lehmann, whose publications he later read in preparation for his inspection journeys, and Andreas Sigismund Marggraf, who since 1753 was director of the new laboratory of the Academy of Sciences. As the Medical-Surgical College did not have the right to confer doctorates, Gerhard continued his medical studies at the University of Frankfurt (Oder), where in 1760 he gained his doctorate with a medical-mineralogical dissertation. He then returned to Berlin and began to practice medicine and publish on medical subjects. From 1762 on, he offered private lectures on physics and mineralogy, through which he gradually acquired a reputation as a man of science.

In 1768 the General Directory established the Mining and Smelting Works Department and appointed the reform-minded Minister vom Hagen as its director.[2] When in the spring of 1768 Hagen was seeking a scientifically educated mining councilor for his new department, Gerhard was the right man for the job. The minister arranged without delay Gerhard's double appointment as mining councilor and member of the Royal Prussian Academy of Sciences. As we have seen in chapter 1, the General Directory and the Academy of Sciences had an excellent relationship, and the salary (*pension*) connected with Gerhard's membership in the Academy initially secured his income.[3] After the establishment of the General Building

FIGURE 7.1
Portrait of Carl Abraham Gerhard.
Source: Stiftung Stadtmuseum Berlin.

Department in June 1770, Gerhard was also appointed a building councilor (*Oberbaurat*) and from that point on drew an annual official's salary. After his appointment as Privy Finance, War and Domain Councilor (*Geheimer Oberfinanz-, Kriegs- and Domänenrat*), the highest title a commoner could attain in the eighteenth-century Prussian state administration, his salary was many times over that of a university professor.[4]

As we have seen in previous chapters, Gerhard was a member of state evaluation commissions, installed lightning rods on the powder tower of the artillery, and invented pigments for the Royal Prussian Porcelain Manufactory. But his main field of technical practice was mining. As a leading official in the central Prussian mining administration, he undertook months-long inspection tours in summer and developed proposals for technical improvements. He was responsible, in particular, for the ironworks in the central parts of Prussia (Kurmark, Neumark), and he developed a particular expertise for work with cast iron, which was of great importance for the

Prussian army. He also occasionally cooperated directly with the Prussian artillery. In 1775 and 1776, for example, he tested the casting of iron cannons at the iron foundry in Vietz (near Küstrin, Neumark) and in the gun factory in Spandau, near Berlin. He also participated in the improvement of iron production in Silesia, which would become the most important producer of iron in the German states in the following decades.[5]

Parallel to his work as a mining official, Gerhard began to publish on mineralogical, geological, and chemical topics. Between 1773 and 1797, he published four books (in six volumes) as well as numerous articles in Crell's *Annalen der Chemie* and the *Mémoires* of the Royal Prussian Academy of Sciences, the last one issued in 1816/1817. In addition to his membership in the Academy of Sciences, he was also elected as a member of the Academia Naturae Curiosorum ("Leopoldina," from 1770), the Munich Academy of Sciences (from 1783), the Prague Academy of Sciences, and several other local scientific and economic societies. In the laboratory of the Mining and Smelting Works Department, which he had set up, he conducted systematic experiments on steel, alloys, and other useful materials. Furthermore, he elaborated a program for the teaching of the mining sciences, was the organizer of the mining administration's series of lectures on the useful sciences, and was also a teacher in the framework of these lectures.

Thus, Gerhard was at once a man of science, a teacher of the useful sciences, a mining expert, and a technological experimenter. Cleary, he exemplifies the figure of a scientific-technological expert in late eighteenth-century Prussia. The interdependence of practical work, technological inquiry, and scientific activities stands out particularly in his case. At the time of his appointment as a mining councilor and member of the Academy of Sciences, apart from his mineralogical dissertation in medicine, Gerhard had not yet published a single paper concerned with the natural sciences. Only after he became a mining councilor did he publish on chemistry, mineralogy, and geology, and gain an international reputation as a man of science. Moreover, although the medical doctor was knowledgeable in mineralogy, in 1768 he was not yet a technical expert on mining. He had neither been apprenticed to a mining official nor had he studied at a mining academy, as Alexander von Humboldt and Dietrich Ludwig G. Karsten would do a few decades later. Beginning in the autumn of 1768, however, Gerhard had ample opportunity to visit mines, coal pits, iron foundries, and metallurgical plants during his annual inspection trips.

Gerhard's appointment as a mining councilor in 1768 was in every respect a turning point in life: he became a technical expert in mining as well as a man of science with a great international reputation. In this chapter we will recount his first inspection tours in Silesia, his homeland. We will see what a mining councilor did on his inspections and how he reported his observations to the central mining administration in Berlin. The strong interdependence of Gerhard's professional practical work and scientific studies are well documented in his reports. Gerhard wrote inspection reports for the mining administration, and at the same time he reported his observations to the Academy of Sciences. In his first chemical-mineralogical book, published in 1773, he explicitly acknowledged the importance of his activities as a mining official for his mineralogical and geological studies. "As my present professional business makes me better acquainted every day with the mineral kingdom, and provides the opportunity for different observations of the bodies belonging to it," he wrote, "I [have] used some of my leisure time to publish these [observations]."[6]

GERHARD'S FIRST INSPECTION IN SILESIA

From September to November 1768, Gerhard undertook his first inspection trip to Silesia, accompanied by the older Privy Finance Councilor (*Geheimer Finanzrat*) Heinrich Wilhelm Reichardt.[7] Indeed, Minister vom Hagen's wish to organize an inspection of Silesia's mining was the immediate motivation for Gerhard's appointment as a mining councilor. On August 2, 1768, a few days after his appointment, Gerhard received 400 Reichsthalers from the Academy of Sciences to cover his travel expenses, and shortly afterward the "committee" set off.[8]

After the end of the Seven Years' War and the final annexation of Silesia, Minister vom Hagen was preparing a new mining law for Silesia. Therefore, he wanted to get an overview of the state of mining in this old mining area. Initially Friedrich II had regarded Silesia as a backward territory, which would supply the older provinces of Prussia with raw materials. But this policy changed in the course of the three Silesian Wars. In the 1750s, upper Silesia became the center of the Prussian iron industry, thereby relieving the country's dependency on iron imports from Sweden. Between 1753 and 1756, the royal iron foundries on the Malapane, a tributary of the Oder river, and the Kreuzburg foundry were established. They produced ammunition and bombs for the artillery as well as cast-iron objects.[9] Along with the mines and the work of the mining agencies, the evaluation of these new ironworks was a major part of Gerhard's inspection program.

Gerhard and Reichardt arrived in Silesia in early September. They were anything but pleased with the willingness of the president of the Silesian chambers, Ernst Wilhelm von Schlabrendorff, and the local Silesian officials to cooperate with the Prussian authorities. The Silesians' *"Schlendrian"* (laziness and inefficiency), a frequent complaint in the months and years to follow, was cause for outrage. The two Prussian inspectors had hoped to have access to documents in the Silesian chambers that would facilitate their inspection. But such evidence of an efficient bureaucracy either did not exist or could not be found, and existing reports about the state of mining, which Schlabren-dorff immediately requested from local officials, arrived only sparingly.

Notwithstanding these difficulties, Gerhard and Reichardt visited the royal ironworks in Malapane and in Kreuzburg. Gerhard's report on them, completed on October 30, 1768, was very critical.[10] In both foundries, the iron ore was insufficiently washed and pounded. Moreover, in Malapane the iron ore was not roasted. Consequently, the ore contained too much sulfur and therefore the crude iron was of inferior quality. Gerhard pointed out further that the iron ores used in Malapane and Kreuzburg contained zinc, which would have been a great commercial advantage if it were separated from the slag. But since this was not the case, he explained, the precious zinc was lost. A third point of criticism was the loss of iron during its refinement; a fourth, that the bellows for the furnaces were not strong enough—and so on. Gerhard's criticism was based not least on the results of his chemical analyses of the iron ores and smelted products. The young mining coun-cilor thus began, on site and through learning by doing, to gain competence in mining technology. But his reporting would hardly have been possible without preparatory reading. The identification of technical flaws required having an overview of the existing possibilities of smelting technology and standards of comparison, which Gerhard could only have acquired through the reading of the contemporary technological literature.[11]

Another purpose of the first inspection trip was to search for unexploited ore deposits. Gerhard tried to obtain as much information as possible about such places from local miners and inhabitants, as well as by reading the literature on Silesian mining that he found in local libraries. Equipped with such information, he traveled in the vicinity of the town of Reichenstein. On November 3, he wrote to Reichardt, who had stayed back in Breslau: "Thank heaven, our wishes are fulfilled, I found very beautiful ore yester-day and today."[12] With the help of a local foreman and a young miner (*Bergknappe*), he had found copper ore in the dumps of an abandoned mine. As the region seemed promising, he also prospected for new deposits on the neighboring hills, but soon gave up as he lacked appropriate instruments.

During his first inspection tour, Gerhard found a number of useful min-erals, among them coal, galena, white spar, as well as yellow mica, which was often associated with tin ore deposits. He identified these minerals on the basis of his mineralogical and chemical knowledge, and collected other minerals for his own mineral collection. But as winter was approaching, the two commissioners had to travel back to Berlin. They agreed that it would be worthwhile to continue their search for ore deposits the next spring, and that they would need better instruments for this purpose.

THE SECOND INSPECTION TRIP: PROSPECTING

Based on Gerhard and Reichardt's report, Minister vom Hagen sent his own report to the king in January 1769, which highlighted the success of the first trip and suggested a second one. Hagen also asked for more money (1,800 Reichsthalers) to buy instruments, among them a mountain drill, mine surveying instruments, and a portable assay furnace. After Friedrich II approved these requests, Gerhard received instructions to buy the instruments, or order their construction, and prepare for the second inspection tour.[13]

The second trip, which primarily served to find new ore deposits, took place from May to October 1769. In preparation, Gerhard immersed himself in the writings of his predecessor, Johann Gottlob Lehmann. Lehmann had published an article concerned with "descriptions of the earth and travels" (*Erd- und Reisebeschreibungen*), which included advice for traveling naturalists.[14] It was both a practical travel guide and a mineralogical-geognostic study that aimed to prepare the traveler for mineralogical observation and further explained structural geognostic relationships. Lehmann intended to establish a "subterranean geography" (*geographia subterranea*) that would provide knowledge about the invisible "things hidden in the earth," that is, along with the minerals, most importantly the rock strata and the spatial orientation of the ore veins.[15] But his subterranean geography also encompassed the topography of the landscape and further issues that later would belong to the field of geology. Among the methods of subterranean geography, Lehmann mentioned chemical assaying and the use of a mountain drill, which was still unusual at the time. Moreover, like Werner after him, he understood subterranean geography as a dual endeavor, which served both natural inquiry into the earth's crust and the provision of reliable useful knowledge about mineral deposits. All of this must have been very stimulating for Gerhard, who frequently mentioned Lehmann in his travel report.

In preparation for the second journey, the committee was not only better equipped with instruments and knowledge but also enlarged by seven mineworkers, four foremen, and an experienced mining master (*Oberbergmeister*) named Elster from the mining town of Zellerfeld in the Harz region. The fact that a foreign mining expert was invited shows that the Prussian mining authorities did not yet have enough mining experts in their employ. In the five months that followed, Gerhard and Elster were rarely separated. They would prospect for useful minerals and visit mines and smelting works almost always together. In this way, Gerhard enjoyed a kind of personal training similar to the traditional apprenticeship of prospective mining officials in the advanced mining regions of Saxony and the Harz region.

From May 16 to October 20, 1769, Gerhard and the mining master Elster prospected on dozens of Silesian hills, examined closed mines and slag heaps, entered pits and galleries, and tested samples of minerals from all of these sites.[16] As a rule, they first sent a foreman and one or two miners to an abandoned mine for a rough check. As many mines were in bad condition, more than once the miners first had to drain a gallery, repair a shaft, and clear it of boulders before they could enter. After these preparations, Gerhard and Elster visited the mine and took ore samples. They inspected the samples' observable features, then performed chemical assays of them using their portable smelting furnace. After the identification of the mineralogical species, further quantitative testing was crucial for deciding whether the mineral could be extracted profitably on an industrial scale.

In addition to examining abandoned mines, Gerhard and Elster also prospected for new deposits, mostly at places they had noted during their first journey. Using their mountain drill, they drilled boreholes, sometimes more than ten meters deep, to take material samples. They then performed assays to identify the minerals that they contained. In some cases they stayed at a place for just one or two days, in other cases for more than a week. Through their qualitative and quantitative chemical tests, Gerhard and Elster found copper, lead, silver, tin, as well as arsenic in the first couple of weeks. In the weeks that followed, they also found coal, iron, alum, calamine, chrysoprase, vitriol, and precious stones.

One particular material the two men were interested in was "cobalt" (*Kobalt*). "Cobalt" (today, cobalt ore) was the raw material for the production of a blue pigment, sold under the name smalt, or cobalt blue (*Blaufarbe*), Saxon blue (*Sächsisch Blau*), and so on, which was used in the Silesian linen industry as well as to color glass and ceramics. Because Prussia spent about 57,000 Reichsthalers annually on imports of Saxon blue, the state was pursuing avenues to replace the import with a domestic product.[17] Shortly before the outbreak of the Seven Years' War, Johann Gottlob Lehmann had prospected for cobalt in Silesia, yet without success. After the war, in 1766, the cameralist writer Johann Heinrich Gottlob Justi (1720–1771), at the time a Prussian official, headed a committee that also searched for Silesian cobalt. The committee failed, however, in part because one of the alleged experts was an imposter.[18] By contrast, on their tour Gerhard and Elster found rich veins of cobalt ore in a mine near the village of Querbach. In order to identify its quality, they submitted samples to experimental tests, which showed that they yielded good sorts of smalt. Gerhard and Elster were also successful in getting a Silesian landowner, Count von Schaffgotsch, to open a cobalt mine. A couple of years later, this mine would be supplemented by a workshop for the manufacture of smalt.[19]

On his second inspection trip, Gerhard thus learned prospecting from the ground up. Under the tutelage of the Harz mining master Elster he became acquainted with techniques of observation and manual skills of test digging, drilling, and ore assaying. After more than five months of prospecting and further mine inspections, the commission returned to Berlin, and on November 10, 1769, Minister vom Hagen received Gerhard and Elster's final report. It contained information about mineral deposits, as well as a complete list of all the mines and smelting works in operation in Silesia, along with an economic balance sheet.[20] It also made proposals for attracting new investors—such as public advertisements and financial benefits for investors—that had long been in the inventory of mercantilist policy. And it made yet another proposal: the establishment of a mining school along with a laboratory in Berlin. During their inspections, Gerhard and Reichardt had met many incompetent officials. If this was to change, they pointed out, Prussia had to create an institution for the education of its mining officials. On December 3, 1769, Minister vom Hagen sent a summary report to the king, in which he identified the lack of knowledgeable officials as the main cause of the deplorable state of mining in Silesia.

GERHARD'S REPORT TO THE ACADEMY OF SCIENCES

During his two inspection tours over the course of a total of eight months, a new world of practice opened up to Gerhard. But the academically trained physician and mining councilor also improved his natural knowledge. While he was still in Silesia, he sent the members of the Academy of Sciences a letter with the newest "natural historical" information from Silesia, which the mathematician Jean Castillon read out at the members' meeting on September 28, 1769. Right after his return to Berlin, Gerhard gave a lecture and personally presented the minerals he had brought back.[21] On his later inspection journeys, Gerhard always was on the lookout for new minerals, so that by 1780 his collection had grown to almost four thousand mineral specimens.

A week after his first academic lecture, Gerhard presented another mineralogical report to the academicians, followed in February 1770 by a geological report, which was published in 1771 in the Academy's *Mémoires*
under the title *Observations physiques et minéralogiques sur les montagnes de
la Silésie*.[22] In this report, Gerhard gave, for the first time, a comprehensive
geological description of the Silesian mining regions, which ranged from
the topography of the landscape to physical measurements and geognostic
considerations on the "types of rock masses" (*Gebirgsarten*) and rock structures. The academic audience was set in the mood for a scientific report with
Gerhard's opening remarks: "The voyage to the Silesian mountains, which
I had the honor of undertaking on behalf of the king, provided me with
the opportunity, Gentlemen, to communicate to you several remarkable
things."[23] What sounded like a research voyage was in fact an inspection
tour to a mining region under the authority of the Mining and Smelting
Works Department.

In his *De re metallica*, Agricola had already recommended that the opera-
tors of mines thoroughly review the contours and composition of the
earth's surface in the region as well as the rivers and the roads.[24] In the eighteenth century, inspection reports often included elaborate topographical
descriptions.[25] Knowledge of the elevation and course of mountain ranges,
the orientation of valleys and rivers, the size of forests, and the landscape as
a whole belonged to the repertoire of useful knowledge that flowed into the
planning and decision making of the mining authorities. If there were no
forests, then there was no fuel; if there were no rivers, then neither wood,
nor ore, nor metals could be transported. The mining officials' inspection
reports and the scientific travel reports thus shared many characteristics.

Gerhard's academic *Mémoire* began with a description of the landscape,
which first addressed the orientation of the mountain ranges and the
shape and height of the mountains and then went into a description of
the lakes and river courses. Data from barometric elevation measurements
and measurements of the air temperature were integrated into this written
description. The main focus of the treatise was, however, on geognostic
issues, including the description and classification of the Silesian "*Gebirge*,"
by which Gerhard, like Lehmann and later Werner, meant large-scale rock
masses.[26] References to other naturalists and natural philosophers, among
them Lehmann, Leibniz, and Buffon, demonstrate that Gerhard compared
his personal observations to collective knowledge, gained through his reading of the relevant literature.

Adopting Lehmann's distinction between *Ganggebirge* (veined rock masses)
and *Flötzgebirge* (layered rock masses), Gerhard differentiated between three

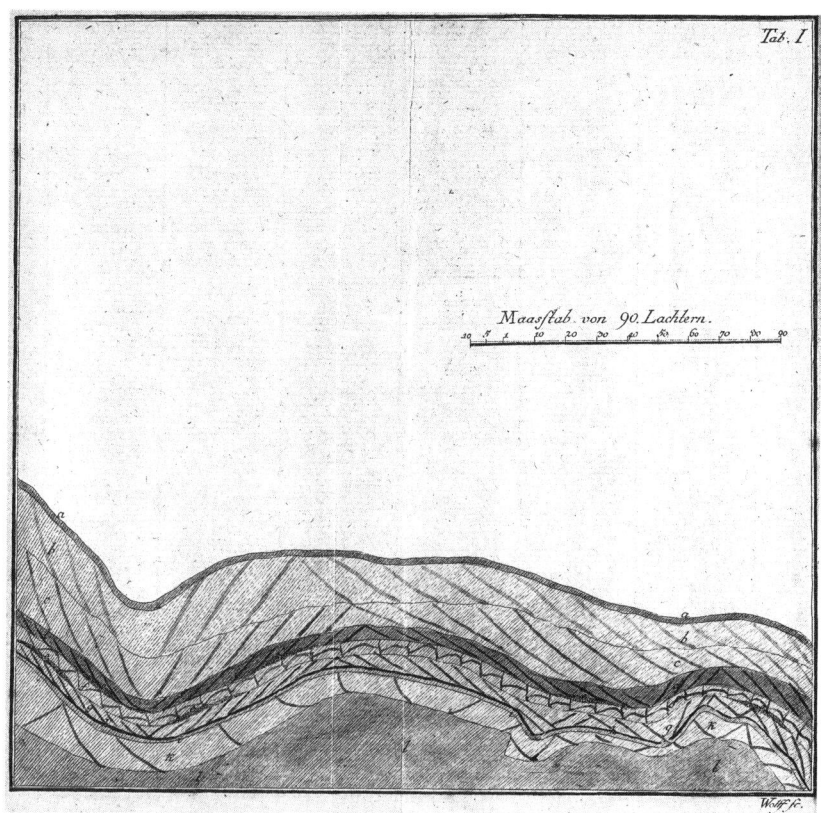

FIGURE 7.2
Diagram by Gerhard: rock strata of *Flötzgebirge*.
Source: Gerhard 1781–1782, vol. 1, plate 1.

types of rock masses, the first being the *Ganggebirge*, which were made up of hard rock, such as granite and gneiss, and contained clearly visible ore gangues that ran through the rock like veins. Following these oldest "primitive rock masses" were the secondary *Flötzgebirge*, which were composed of laterally deposited layers of softer types of rock, such as sandstone and limestone, in which useful minerals, including hard coal, were laterally deposited. The third type of rock mass formed the transition from veined to layered rock masses and was classified historically-genetically as between the oldest primitive rock masses and the younger *Flötzgebirge*. Later, Werner made similar classifications.

Like Lehmann, Gerhard also made a correlation between the "structure of the rock masses" (*structure de montagne*) on the one hand and the type of useful minerals contained therein on the other. This kind of knowledge, he stated, would be "of great utility for the miners."[27] Gerhard's concept of *structure de montagne* is consistent with Werner's later concept of rock formation (*Gesteinsformation*), discussed in the previous chapter, as it also was based on historical-genetic considerations.[28] The metal veins contained in rocks, Gerhard pointed out, differed in their extension and the depth to which they descended. Gerhard linked this observation with the hypothesis that the formation of mineral veins presupposed fissures of rocks caused mainly by desiccation following floods. As a man of the Enlightenment, he did not rely on the Bible, as Lehmann had done, and rejected the idea of the Noachian flood. Among other things, he put forth the argument that a single, long-lasting flooding of the earth would have led to a more regular pattern of mineral deposits, corresponding to the minerals' weights. Instead, he conjectured that there were multiple, alternating periods of flooding and retreat of water through which fissures in the rock masses formed and later, enriched with minerals, appeared as veins.

Gerhard's *Mémoire* contains descriptions of minerals, discussion of geognostic topics, as well as a topographical description of the landscape, that is, what he later termed "physical geography" or *geographia physica*.[29] Thus, it combined new approaches in the studies of the earth of the time that constituted the new discipline of geology. It also presented a wealth of information about the usefulness of Silesian minerals. For example, Gerhard informed the Academy members about his discovery of cobalt ores and the opening of a cobalt mine by Count von Schaffgotsch.[30]

In his later textbook, *Essay on the History of the Mineral Kingdom* (*Versuch einer Geschichte des Mineralreichs*), Gerhard dealt again, this time more systematically, with geological subjects. Published several years before Werner's geognostic main work, *Short Classification and Description of the Different Kinds of Rocks* (*Kurze Klassification und Beschreibung der verschiedenen Gebirgsarten*), this textbook dealt with the mineral kingdom in a comprehensive way and thus included classical mineralogy, *geographia physica*, the study of "*Gebirge*" in the sense of large-scale rock masses, that is, geognosy, and the study of ore deposits.[31] Written for Gerhard's teaching in the context of the lecture series of the Berlin mining administration, it was aimed at future mining officials. In the learned world it was also greeted emphatically as a pathbreaking work on the latest earth research.[32]

Historians of geology largely agree that mining was a significant context for the formation of the new science of geology in the decades around 1800. The question of whether mining had an impact on all fields of geology, however, is still a controversial issue. While the argument that observations in mines contributed to geognostic knowledge about subterranean rock strata is compelling, it is perhaps less plausible to assume that mining and mining academies promoted outdoor fieldwork and studies of the physical landscape. The historian of geology Martin Rudwick regards physical geography—defined as the systematic study of the characteristics of the earth's surface—as an endeavor pursued by "gentlemen scientists" and, as a rule, independently of the consideration of its utility. "Physical geography belonged to the world of cultured travels and regional surveys of natural and human resources," he writes, "the world of savants such as Saussure and Hamilton. Geognosy had a much more specific home, in the world of mining."[33] Yet, as we have seen, Gerhard did include physical geography in his *Mémoire* from 1771, and in his later writings, which undeniably emerged in the context of mining. The same applies to Lehmann. Moreover, as evidenced in Gerhard's archived reports to the mining administration, he also included "physical geography" in his lectures for the mining officials.[34] For mining officials such as Lehmann, Gerhard, and, as we will see in chapter 10, Alexander von Humboldt, both geognosy and physical geography were potentially useful knowledge.

8 EXPERIMENTS IN THE LABORATORY OF THE MINING DEPARTMENT

We have already mentioned that in their inspection report on Silesia of November 1769, Gerhard and Elster proposed the establishment of a mining school in Berlin along with a laboratory. In fall of 1770, in place of an actual mining school, the Mining and Smelting Works Department started a lecture series on the useful sciences. When in summer of 1774 Minister Waitz von Eschen became the new head of the department, he also convinced the king that the mining department needed a laboratory for the assaying of ores and for teaching.

Shortly thereafter, Gerhard leased a small outbuilding in the courtyard of his residence on Boulevard Unter den Linden and transformed its kitchen into a laboratory.[1] In November of the same year, he began to offer experimental metallurgical-chemical lessons on Saturday afternoons.[2] In the years that followed, he regularly used the laboratory in winter for teaching as well as for technological experiments on materials. Like his inspection reports in the summer, he reported these experiments both to the mining administration and the Royal Prussian Academy of Sciences.

In October 1775, Gerhard presented a first report on his experimental lessons to the minister, writing that in his *"Cursum Chymicum"* he had conducted altogether one hundred eight experiments in the presence of his students and documented them in a "journal." The experiments dealt with the following chemical-technological topics: the investigation of the properties of iron and steel; the analysis of ores of silver, copper, lead, iron, and mercury; the analysis of cobalt ores and the subsequent investigation of the nature of the blue pigment in cobalt; and the production of crystal and lead glass as well as colored glass. Gerhard announced that he wished to carry out similar experimental investigations in the near future, because they would have a great influence on industry.[3]

One week later, he submitted a detailed plan for his subsequent experiments, along with a cost estimate, to Minister Waitz von Eschen for approval.[4] He intended to first conduct a series of experiments of relevance to the production of glass and ceramics. In a first series of experiments, he aimed to produce materials made of two components. To this end, he wished to heat different mixtures of one type of earth and one type of rock in a wind furnace, and test whether a homogeneous mass could be produced this way. Using altogether thirty different rock and earth types, the first experimental series would comprise nine hundred individual experiments. A second series of experiments would follow, in which three different types of earths were to be melted into a homogeneous material. In a third series, metalliferous admixtures would be added to test their influence on the melting behavior of the materials. Gerhard estimated that all three series of experiments would take one and a half years to complete, distributed over three successive winters.

The second part of his plan proposed an experiment on the fusion of metals, to investigate possibilities for the production of new metallic alloys. Bronze and brass, the alloys of copper and tin and copper and zinc known since antiquity, were already well-investigated materials that possessed different physical properties and practical applications than their metallic components. Therefore, Gerhard wanted to examine the behavior of alloys of pig iron with copper, tin, and zinc as well as the alloy of copper and bismuth. After the preparation of new alloys, tests of their properties would follow, including their specific weight, tear resistance, brittleness, deformability, color, and metallic shine. Minister Waitz von Eschen approved these experiments and made funds available for their commencement in the winter of 1775–1776.

COBALT, IRON, AND STEEL

In an essay on cobalt (*Kobalt*), published in the *Mémoires* of the Academy of Sciences in 1779, Gerhard reported the first results of his experiments. With these experiments, he was at the forefront of chemical research. As we have already seen, cobalt was the name for cobalt ore, which was processed into blue pigments such as cobalt blue. Late eighteenth-century chemists agreed that cobalt was a mixture of various substances, but there was no consensus on its composition. Furthermore, it was unclear which components of cobalt caused the blue color. In 1735, in a treatise for the Swedish Academy of Sciences, the Swedish chemist Georg Brandt had asserted that he had found a metallic regulus in the chemical analysis of cobalt.[5] He can thus be regarded as the discoverer of the metal that today we call cobalt, but until the end of the century, his discovery was questioned or even ignored. For example, Gerhard's predecessor Johan Gottlob Lehmann had dedicated a two-volume treatise to cobalt, *Cadmologia oder Geschichte des Farbencobalts* (1760), in which he contradicted Brandt's view that cobalt contained a new kind of metal.

In his chemical analyses of cobalt, which comprised several stages, Gerhard succeeded in its complete decomposition and in identifying its components. He proved that the raw material cobalt consisted of various earths and metals. Moreover, he identified a metal that he designated as the "true metallic cobalt" and the "true coloring metal," thereby confirming Brandt's earlier discovery. At the same time, he claimed to have invented the "true method" of producing the purest blue pigment, called *Königsblau*. His pigment, he observed, could be used for the decoration of porcelain.[6] Here discovery and invention were intimately connected.

Gerhard's second report, published in 1780 in the *Mémoires* of the Academy of Sciences, was concerned with iron and steel.[7] The report focuses on the technical problems facing iron smelting as well as the different variants of iron and steel. In accordance with its mercantilist policy, Prussia had long attempted to reduce imports of Swedish iron, and to replace it through domestic products. Gerhard was directly tasked with this goal, as he had the oversight of the iron foundries in Neumark, Kurmark, and Pomerania. Between 1775 and 1776, he had participated in trials of the casting of cannons in the iron foundries in Vietz, near Küstrin, and in the arms manufactory in Spandau.[8] Hence, his experiments with iron and steel and his publication in 1780 stood in a larger economic context. In his report, Gerhard described problems of iron smelting, but he was not able to propose solution to these problems.

Another topic was the still unsolved problem of chemical identification of the various sorts of iron and steel. The commercial varieties of iron and steel presented a great challenge to the eighteenth-century chemists. On the one hand, from a chemical perspective, all these varieties counted as a single kind of chemical substance, that is, iron. The reason for this was their identical reaction with mineral acids, which transformed them into the same kinds of iron salts. On the other hand, from a practical perspective the various properties of the commercial varieties of iron and steel were of great significance. Chemists assumed that the variability of iron was caused by small differences in chemical composition. However, quantitative chemical analyses were not precise enough to confirm this assumption. Again, Gerhard was not able to report experimental success in this respect.

GLASS AND CERAMICS

In 1781, Gerhard published two more essays in the *Mémoires* of the Academy of Sciences, which again dealt with earths and rocks, as well as with the fabrication of glass and ceramics. The first publication focused on a purely technical problem. Gerhard had established that the results of his smelting experiments with earth and rock types depended on the type of melting pot he used, and therefore were not always reproducible or comparable to the test results of other chemists. Hence, he conducted a systematic comparison of the melting behavior of ninety-eight different types of earth and rock in three different types of pot. He summarized the results of these experiments in a fifteen-page table, while at the same time he made reference to the usefulness of his experiments for industrial metal smelting and for the production of porcelain and ceramics.[9]

In the second publication from 1781, Gerhard compiled the results of his melting experiments on mixtures of an earth and a rock type, which he had conducted some years earlier. The emphasis in this text now lay on the technical goal of the improvement of glass production and the invention of new types of glass. The first experiments dealt with the invention of a durable glass for everyday use. Gerhard was looking for a replacement material for potash, which he believed was detrimental to the strength and transparency of glass. In melting experiments with feldspar and chalk, he obtained a hard, greenish glass, which, however, did not have a smooth surface. In contrast, the different mixture and melting experiments with quartz, feldspar, and chalk did produce a new kind of "perfect glass." To make the production of this glass economically viable, in subsequent trials Gerhard replaced the chalk with cheaper lime "from Rudersdorff." Here too, he was successful and announced that he had done a first step toward the invention of a newer, harder type of glass. All things considered, he observed, his new type of glass was even cheaper to produce than the conventional glass produced with quartz sand and potash. But success with the experiments on the production of crystal and colored glass still eluded him. He thus announced that he wished to continue with them.[10]

CONTINUATION OF THE EXPERIMENTS ON EARTHS, ROCKS, AND ALLOYS

In January 1783, Gerhard submitted a proposal for "useful work in the laboratory," which he had prepared for Minister von Heinitz. In this paper, he distinguished between three functions of the mining administration's laboratory: it was a place for the instruction of mining students (*Bergeleven*), for the assaying of ores, and for research, which should lead to "new and useful discoveries and inventions."[11] The laboratory instruction for the mining students, Gerhard further pointed out, would henceforth be held Wednesdays and Saturdays, to be taught mainly by the mining assayer Weiland, to whom he would provide the necessary instructions. Weiland had been hired in the fall of 1782, after repeated requests from Gerhard for help on account of a work overload. Alongside the support for Gerhard in experimentation and teaching, he was also supposed to practice his actual profession as an assayer. The latter depended on when the ores from the mining regions were delivered, but when possible, assaying was to be carried out also on Wednesdays and Saturdays and in the presence of the students. "All experiments must be recorded in a special book," emphasized Gerhard.

Gerhard further proposed that the other four weekdays would be reserved for his experimental research on earths, rocks, and alloys. The "most skilled, eldest mining students" would be admitted to the experiments. Weiland was supposed to perform these experiments and pick up the recipes for them beforehand. Gerhard expected that each evening he would report the experimental results and further keep a double protocol. After one week, the notebooks were to be presented to Gerhard for his review.[12] The expectations of the mining assayer were thus high, in particular with a view to the written reporting. As we have seen above, written reporting was a request of the Mining and Smelting Works Department under the direction of Heinitz.

As regards his experimental research, Gerhard intended first of all a "continuation of the steel-melting experiments" that he had already mentioned in his publication on iron in 1780. Primarily, he wished to investigate the question of whether there were material additives that increased steel density. Following these experiments, he intended to turn again to the "glass experiments," focusing now on the production of crystal and colored glass. He reported that in his earlier experiments he already had some success, but for the transfer of his techniques to a technical scale he was still lacking important details. The production of new alloys was a further goal. Referring to his experiments carried out a few years earlier, Gerhard pointed out that he had already "created alloys of various metals for all kinds of use" but had not exhausted all possibilities. There were still "very useful discoveries" to be made, and he estimated that it would take about one thousand experiments to achieve his goal.[13] Around the same time, Gerhard's friend Franz Carl Achard undertook similar experiments in the laboratory of the Academy of Sciences, presumably at Gerhard's suggestion.

Friedrich II approved the majority of the proposals, while not forgetting to point out that Gerhard should take care to "make economical use of the provided materials." For reasons of cost, however, he refused to approve the metallurgical experiments. "We have resolved to suspend the experiments that you have proposed for the discovery of new metal compositions," he wrote, "because, due to the quantity and reguli they require, they are too costly and too lengthy."[14] In the end, Gerhard was not able to complete these experiments, but for entirely different reasons. In a letter to his longtime correspondent, the Viennese mineralogist Nicolas Joseph Jacquin, he therefore only mentioned the success of his experiments with glass, remarking that "cobalt quartz," when mixed with alkali, produced a beautiful blue glass.[15]

Gerhard was not on the best of terms with Minister von Heinitz, who had entered Prussian state service in the fall of 1777. Only a few weeks after the death of Friedrich II on August 18, 1786, Heinitz made arrangements that forced Gerhard to end his teaching activities.[16] A short time later, he had to give up the laboratory as well. On June 18, 1787, while still on an inspection tour, he received a brusque letter from Friedrich Wilhelm II in which he was ordered to immediately vacate the laboratory. Parallel to this, Heinitz invited the Swedish mineralogist Johann Jacob Ferber to enter Prussian service and become the director of a newly built laboratory. Heinitz wanted Ferber to carry out experiments on the amalgamation of gold and silver on the basis of a new method that had been invented by Ignaz von Born and was being applied in the Saxon amalgamation plant in Halsbrücke. The minister feared that Prussia might miss out on one of the newest technological methods of metal production. Ferber was acquainted with these methods, as he had witnessed them personally. In summer of 1787 construction began on the new laboratory, located in the building of the old mint. But the main part of Heinitz's plan did not come to fruition. In September 1789 Ferber got seriously ill, and in April 1790 he died. His successor was the young mineralogist Dietrich Ludwig Gustav Karsten, who had long been Heinitz's protégé.

After Gerhard had lost his teaching authorization and his laboratory, the Royal Prussian Academy of Sciences became the main institution where he could stay scientifically active. In 1797, he published his last comprehensive mineralogical work, the *Outline of a New Mineral System* (*Grundriß eines neuen Mineralsystems*), for which he revised once more his mineralogical classification system.[17] When in 1810 the seventy-two-year-old was pensioned against his will on account of a loyalty oath to Napoleon, he took it as an affront. Despite the numerous setbacks and intrigues the commoner at the top of the Prussian mining administration had to endure, he was still strongly attached to the cameralist ideal of the expert official promoting useful knowledge, technological progress, and the common good.

9 THE LECTURE SERIES OF THE MINING ADMINISTRATION

In the historical literature, it has long been claimed that a mining academy was founded in Berlin in 1770. Yet we will see in this chapter that this was not the case. Instead, the Department of Mining and Smelting Works instructed Gerhard to organize a series of lectures concerned with mining science and the useful sciences more broadly. The intended audience was to be the future officials of the administrations of mining, building, and forestry.[1] When in 1777 Heinitz became head of the Prussian mining administration, the lecture subjects were strictly limited to mining science. With the founding of the University of Berlin in 1810, the series was scrapped and parts of the lecture series integrated into the university curriculum. Only in 1860 was a real mining academy, comparable to the mining academies in Freiberg or Schemnitz, established in Berlin. The claim that a mining academy was founded in Berlin almost a century earlier goes back to Paul Krusch.[2] In 1904 Krusch, at the time director of the Berlin Mining Academy, published a booklet on the history of his institution. In this publication, he created a myth of origin, presumably for policy reasons, that adorned his institution with the blessings of Friedrich II. Ironically, Friedrich II had been anything but a supporter of a Prussian mining academy.

Between January and April 1770, the reform-minded ministers in the General Directory had indeed undertaken a number of initiatives to found a Prussian mining school. In early January 1770, Minster vom Hagen asked Gerhard to elaborate a detailed plan for such an institution. A few weeks later, Gerhard had drafted the plan, which foresaw that all the Prussian mining officials would first receive two and a half years of scientific instruction in Berlin, and then one year of practical training in the mining regions (see chapter 11). The king, however, rejected the plan on financial grounds. In a diplomatic tone, he informed his minister that he agreed, in principal, with the plan to educate "proficient subjects," but the foundation of the school needed to be "postponed until the next annual financial budget on account of its costs."[3] Hagen understood what this meant and immediately went in search of alternatives.

LECTURES ON THE USEFUL SCIENCES

In the middle of February, Hagen wrote to the Privy State and Justice Minister Carl Joseph Maximilian Freiherr von Fürst und Kupferberg (1717–1790) that it was necessary "that, together and from all sides, we finally seriously think about" what to do about the "general uselessness of a great part of the public servants" and with regard to the multitude of "idiots and ignoramuses" in royal service. There were too few officials, he observed, with "a thorough knowledge of physics, especially of the practical part of the same, of mineralogy and metallurgy and the *mathesi applicata*, also forestry, the knowledge of trees, their planting and insemination and cultivation." Hagen was thinking here of knowledgeable officials not only for the Department of Mining and Smelting Works and the mining administrations in the provinces, but also for the Department of Forestry, just founded in January, and the General Department of Building that was in planning. He then proposed that the Academy of Sciences might be a good location to offer lectures on "the practical parts of the sciences," because it was well equipped with a laboratory, a physical cabinet, a botanical garden, and other resources.[4]

In his response, Minister von Fürst und Kupferberg fully agreed that the Prussian state administration needed "more capable and useful servants." Further, he concurred with Hagen's suggestion to concentrate on the education of future officials for the state's technical departments. He thus proposed that they should restrict themselves to the teaching "of those sciences that concern directly the mines, building [civil engineering], and forestry." However, von Fürst und Kupferberg was skeptical that Berlin was really the right place for the teaching of mining science, as it lacked opportunities for practical training. He also rejected the idea that the Academy of Sciences might be a good substitute for a real school. The purpose of the Royal Prussian Academy of Sciences, he argued, was not to teach but to "make discoveries and inventions" and "extend and improve the realm of the sciences and of human knowledge in general."[5]

Worried about von Fürst und Kupferberg's objection concerning the teaching of mining science in Berlin, in March Hagen instructed Gerhard to visit the Freiberg Mining Academy to make inquiries into its professors' and teachers' methods of teaching. Back in Berlin, Gerhard reported about their combination of practice and theory. The students of the Freiberg Mining Academy, he stated, attended lectures on metallurgical chemistry, theoretical mechanics, practical geometry, and drawing, and they learnt the mining technology "*ex usu* [from experience] and during their daily visit of mines."[6] It must have been immediately clear to Hagen that the Freiberg Mining Academy did not furnish him with new arguments, as Berlin was not a location to emulate the practical training at Freiberg. As a result, he finally gave up his original plan of founding a mining school in Berlin. One day later, on March 25, 1770, he checked a four-year-old file that documented Friedrich II's complaints about the "*négligence*" of his civil engineers (*Baubeamte*). In a marginal note on the file he wrote, "in order to effect the royal majesty's approval of the project to advance the practical sciences, we must not multiply the establishments, but rather combine the General Building Department with our already prepared establishments in the Mining Department."[7]

At the time, Hagen was preparing the foundation of the General Building Department, which took place in June of the same year. As he was well aware of the king's interest in architecture, he made this interest the starting point for his strategy to establish a lecture series encompassing the "practical" or "useful sciences." In a proposal, dated April 7, 1770, for the king, he first recommended the establishment of a General Building Department. He then went on to propose the establishment of "practical colleges" (*praktischer Collegia*) for the future officials of the building, mining, and forestry administrations.[8] The king approved this proposal, apparently not realizing that Hagen's proposal concerning the practical colleges went beyond the training of building officials. Whereas the king explicitly consented to lectures just for the latter, Hagen went on to realize his own, broader plans. Two days later, "with great pleasure" he informed his allied ministers that he had gotten the king's approval.[9]

Enabled by Minister vom Hagen's policy, the lecture series grew out of a compromise that secured the useful sciences a place of their own. But the archival sources do not uphold the further assertion that in 1770 an actual mining academy was founded in Berlin. As regards the institutional framework, Hagen left no room for interpretation, clearly stating that the "currently arranged instruction of various sciences" was "*no new institutum*," but rather had the purpose "that young people be instructed in such *useful sciences* that hitherto were so very neglected" and thereby ensure that "the state accrue able subjects."[10]

The lectures began on October 15, 1770, and were announced in newspaper advertisements, which stated that the "advancement of *useful sciences*" was of great importance to the royal majesty, and there would be "no lack of opportunity" for the instruction of all officials who wished to "dedicate themselves to construction [civil engineering], mining, agriculture, forestry, and all other cameral and finance subjects."[11] The advertisement stated further that in the future such officials would be expected to have thorough knowledge of these sciences, and they should thus contact the instructors immediately. For the teachers, the modest annual honorarium of one hundred thalers was foreseen, which also had to cover all teaching expenses.[12]

The latter idea soon turned out to be problematic, as the apothecary-chemist Valentin Rose (1736–1771), invited to teach chemistry in his own pharmaceutical laboratory, complained that the salary would not cover his expenses for experiments. As he was also seriously ill, he declined the invitation.[13] In March 1771, the physician Ernst Gottfried Kurella (1725–1799), stepson of the academic chemist Johann Heinrich Pott, replaced Rose, after he had consented to concentrate his lectures on the technical parts of chemistry and to leave out its medical and pharmaceutical parts.[14] Johann Gottlieb Walter (1734–1818), a professor at the Medical-Surgical College who already possessed "rare and expensive instruments for experimental physics," taught physics. Johann Gottlieb Gleditsch agreed to teach forestry and "practical botany." Mathematics was first taught by the mathematician Friedrich A. M. Castillion (the Younger, 1747–1814) and from spring of 1771 by the building councilor Friedrich Holsche (1743–1783), who extended his courses to include "practical geometry," architecture (*Zivilbaukunst*), mechanics, and hydrostatics.[15] Gerhard, the organizer of the lecture series, was to report regularly on how the lecture series was progressing and request a list of participants, from which the mining department hoped to gain an overview to help expedite the recruitment of good mining officials. Because attendance at the lecture series was voluntary, there was no strict regulation of the training of these officials.

Gerhard was also the lecture series' teacher of mineralogy and metallurgy. As we have seen, from 1774 on, after he had set up a laboratory, he regularly taught experimental courses on metallurgy and the preparation of materials. Hence he reported to the minster that he had "begun to teach mining science."[16] In addition, Gerhard also taught "theoretical" mineralogy, that is, the rules of mineralogical identification and classification. In 1773, he published the first volume of his mineralogical-chemical textbook for use in his lectures, *Contributions to the Chemistry and History of the Mineral Kingdom* (*Beiträge zu Chymie und Geschichte des Mineralreichs*). His aim, he pointed out, was to construct a "natural system" of minerals based on "chemical principles," that is, on knowledge of the chemical composition of minerals. He observed that his teaching had shown that mineralogy helped his students "to identify and classify minerals they have never seen before."[17] The latter corresponded with the didactic goals formulated in his plan for a mining academy, a goal that we will return to in part III.

Like most mineralogists of his time, Gerhard divided minerals into four classes: the earths and rocks, salts, metals, and combustible minerals. But in the second part of his *Beiträge* (1776), he broke new ground. He concentrated in this part on the combustible minerals and a special kind of them: hard coal. The "widespread wood shortage," and the rich coal deposits in Silesia, he pointed out, made "an exact knowledge" of hard coal imperative.[18] Ironworks, which belonged to Gerhard's area of responsibility in the Department of Mining and Smelting Works, were some of the major consumers of wood at the time, and the increasing shortage of wood was a limiting factor of the growth of the iron industry. Therefore, the search for a replacement for wood was a key element in the state's management and promotion of industry. As Joachim Radkau has pointed out, in hindsight these endeavors appear to have been stimuli of a beginning industrialization of Prussia.[19] Over one hundred thirty pages long, Gerhard's text was the most comprehensive treatise on hard coal at that time. In 1779, King Friedrich II granted the Mining and Smelting Works Department permission for the large-scale mining of Silesian coal.[20]

HEINITZ'S REORGANIZATION: THE TEACHING OF MINING SCIENCE

In September 1777, Friedrich Anton von Heinitz had entered into Prussian state service as the minister responsible for the Mining and Smelting Works Department. This appointment entailed the transfer of the reform strategy Heinitz had developed in Saxony to Prussia. Under Heinitz, a thorough restructuring of the lecture series and a strict regulation of the education and training of Prussian mining officials were implemented. First, the new minister restricted the lecture series' participants to mining officials, eliminating the lectures for the candidates of the forestry and building administrations. He apparently disagreed with the existing regulation that had the mining department carry the costs for the education of those other officials. As a result, he immediately dismissed Gleditsch, who was responsible for forestry, and in fall of 1779 Holsche, who had been the link to the General Building Department. Walter and Kurella were also dismissed because they had submitted fake reports on courses they had actually never taught.[21]

Another measure placed a stronger accent on the practical training of the Prussian mining officials. On December 15, 1777, Heinitz issued a directive that stipulated that only those young persons "who are familiar with the local affairs of a mining and smelting works and have already spent some years in mining regions" should "dedicate themselves to the sciences."[22] Here it is plain to see that "practice" was to come before "theory." Thus, Gerhard's previous plan for a mining academy, in which the priorities were reversed, was definitively abandoned. Clearly, Heinitz, the cofounder of the Freiberg Mining Academy, was skeptical about the importance that "theory," that is, lectures, had received in Berlin.

Heinitz's directive also ordered the establishment of local schools for the children of mining and smelting works officials, from whom the corps of mining officials traditionally was recruited. Gerhard received the directive to organize the school curriculum and to ensure that the regional mining offices would receive the "easiest, most thorough, and comprehensible compendia" for this task.[23] In January 1778, Heinitz concretized his reform plans through a "public notification [*Publicandum*] on how future appointments of mining and smelting works personnel should be handled." Each candidate for employment in a Prussian mining agency would have to "completely qualify" himself for his position. He would first have to register with the local mining office in Reichenstein, Rothenburg, or Hagen, and "personally make himself practically acquainted with all the kinds of work and commercial matters." This practical part of the training would last one to two years and be supplemented with classroom instruction in winter. Afterward, the best "trainees" (*Eleven*) would be promoted to "mining cadets" (*Berg-Cadetten*) and then receive a theoretical-academic education, wherever "there is the opportunity for it." Since "opportunities" for higher scientific education were also available at the Prussian universities and at the Freiberg Mining Academy, Heinitz's *Publicandum* yet again rejected a central mining academy in Berlin serving all of Prussia. Already in the fall of 1777, Heinitz had sent four young mining official candidates to study at the Freiberg Mining Academy. To guarantee the free choice of the place of scientific study, the cadets were granted a scholarship. The *Publicandum* also foresaw an examination at the end of the course of study, which would be conducted by the regional head mining offices.[24]

The *Publicandum* of January 1778 introduced for the first time binding regulations for the professional training of Prussian mining officials, and established that this training would have two phases: a practical one with trainee (*Eleven*) status, and a subsequent scientific-academic one with "cadet" status and concluding examinations. However, in practice the course of training could differ, as the career of Alexander von Humboldt shows. As we will see in the next chapter, in 1792 the young nobleman was hired as a Prussian mining assessor after just a brief, nine-month-long attendance at the Freiberg Mining Academy, and without prior trainee or cadet status or an examination.

Heinitz's *Publicandum* also brought about a change in status of the lecture series in Berlin. Hitherto organized and financed by the Mining and Smelting Works Department, which was responsible for all of Prussia, its organization was henceforth transferred to the "Mining and Smelting Administration," which had a regional status similar to that of the mining and smelting offices in the local mining regions. Furthermore, the lecture series' function was limited to the schooling of lower, technical mining officials who worked within the purview of the local mining administrations. If one wished to pursue a career as a higher mining official and become a mining councilor or mining master, additional studies at a university or at the Freiberg Mining Academy were necessary.

As regards the training of technical mining officials, the reorganized lecture series seems to have been a success. Among the best-known students were Karl Friedrich Bückling (1756–1812) and Friedrich August Alexander Eversmann (1759–1837). Bückling attended the lectures in 1773; later he participated in the lecture series as a drawing instructor. He became a famous Prussian mining engineer, organizing the construction of steam engines in saltworks, Silesian ironworks, and the Royal Prussian Porcelain Manufactory. Eversmann was a trainee in Berlin in the early 1780s. In 1783–1784 he traveled to England, and then became a factories commissioner (*Fabrikenkommissar*) in the County of Mark, where he initiated the introduction of coke in iron production and innovations in steel production. He was also a prominent author of scientific-technological texts. In the late 1780s, laboratory workers and arcanists of the Royal Prussian Porcelain Manufactory also attended the mining lectures.[23]

The prestige of the lecture series continued to be secured through the teachers drawn from the ranks of the Academy of Sciences. From 1784, Klaproth lectured on chemistry, Franz Carl Achard taught physics, the head mining and construction councilor Bernhard Friedrich Moennich (1741–1800) taught mathematics and mechanics, and from 1789 the mineralogist and mining assessor Dietrich L.G. Karsten taught mineralogy, after Gerhard was relieved of his teaching duties in the fall of 1786. With the founding of the University of Berlin, the lectures of the mining administration were integrated into the university curriculum, as will be discussed further in chapter 14. The mining students received tuition-free access to the university lectures on chemistry, technology, mineralogy, and physics. Not until 1860, the high point of the first wave of industrialization in Prussia, were the conditions created for a sustained institutionalization of mining science and the founding of a real mining academy in Berlin.

10 MINES AS LABORATORIES: HUMBOLDT

From September 1792 to February 1797, Alexander von Humboldt was the chief mining master (*Oberbergmeister*) in the Franconian parts of Prussia. As we have seen in chapter 6, mining masters were high officials who directed the local mining offices along with the mines and smelting works under their authority. Although they were high officials, they were involved in the daily practice of mining. This chapter will provide an overview of Humboldt's professional activities and detail some of his scientific inquires and inventions. Humboldt regularly visited mines, inspected smelting works, wrote inspection reports, and calculated budgets; he also founded a mining school and invented a miners' lamp and rescue apparatus for miners. His professional activities also provided numerous opportunities for studies of nature. He performed chemical experiments in the mines, used his inspection tours for mineralogical and geognostic observations, published texts on mineralogy, geognosy, chemistry, botany, and physiology, and communicated with men of science all over the world. During his time as a mining master, he became a member of the republic of letters and was admitted to the Leopoldine Carolinian Academy and the Gesellschaft Naturforschender Freunde zu Berlin. Like Achard, Klaproth, Werner, and Gerhard, the young Humboldt was a scientific-technological expert.

On May 14, 1791, the twenty-one-year-old Humboldt wrote a letter to Minister von Heinitz in which he declared that he wished to serve his fatherland as a scientifically trained, practical mining official.[1] He had already briefly studied at university, first from fall of 1787 to spring of 1788 at the University of Frankfurt/Oder where he studied the cameral sciences, and then from April 1789 to March 1790 at the University of Göttingen where he attended, among other things, Beckmann's course on technology (*Technologie*). From summer of 1790 to spring of 1791, he attended the Commercial Academy (Handelsakademie) in Hamburg. During this time, he

Alexander von Humboldt
im 27. Lebensjahre.
(1796.)

FIGURE 10.1
Portrait of the young Alexander von Humboldt by Alfred Krause.
Source: Bruhns 1969, vol. 1, n.p.

also began to study mineralogy, botany, and chemistry. Having decided to become a mining official, he then planned to attend the Freiberg Mining Academy.

On June 14, 1791, Humboldt arrived in Freiberg, where he spent the following eight months. He studied mineralogy and geognosy with Werner, metallurgical chemistry with the seventy-eight-year-old Gellert, and mathematics and physics with Johann Friedrich Lempe (1757–1801). Werner, who also instructed him personally, became his role model. On his daily visits to the mines, Humboldt learned the methods of mining with his own hands and became acquainted with mining technology. He also gained insight into mine surveying, assaying, and technical drawing and wrote reports about his practical work. In addition, he had numerous opportunities to collect minerals, observe rocks and veins, measure airs in mines, and experiment with subterranean plants.[2] Much of Humboldt's knowledge about instruments and skill in measurement, for which he would later become famous, stemmed from his time at the Freiberg Mining Academy.

In late February 1792, Humboldt returned to Berlin to take up his appointment as an assessor at the mining administration. The king had decided, his certificate of appointment read, "to use the knowledge that Alexander von Humboldt has acquired theoretically and practically in mathematics, physics, natural history, chemistry, technology, mining science, metallurgy, and commerce, in the mining and metallurgical service of his highest Majesty."[3] Humboldt was eager to do practical work, and he was "proud to have studied as a practical miner." But the young noblemen also had to defend himself against criticism of his practical-technical interests. Count Friedrich Wilhelm von Reden (1752–1815), a nephew of Heinitz and director of the Silesian mining administration, accused him of having "too small-mindedly studied practical matters." A man of his social standing, he stated, was not born to be a technical official. Little impressed, Humboldt responded that all achievements in mining "depended on the precise study of the technical [des Technischen]."[4]

In Berlin, Humboldt's first duty was to familiarize himself with paperwork, accounting, and the occasional writing of reports on ironworks and chalk pits. In early June 1792, he came a great deal closer to his goal. Heinitz instructed him to travel to the Franconian principalities of Ansbach and Bayreuth, which had just recently become Prussian territories, to inspect the state of the mining industry there.

HUMBOLDT'S FIRST INSPECTION REPORT

In early July 1792, Humboldt arrived in Franconia; Heinitz himself intended to follow in August. The formerly prospering Franconian mining industry was still suffering from the long-term effects of the Thirty Years' War (1618–1648). Mainly it was the extraction of iron that was profitable, and to some extent vitriol as well. In the period from 1769 to 1791 the last margrave, Christian Friedrich Carl Alexander, had issued several regulations that allowed state intervention in mining, although the direction principle was not formally anchored in Franconion mining law. In June 1769, a central mining department was established. Its officials were authorized to regulate "all and every matter of mining," including adjudication and technical improvements.[5] Thus, favorable conditions existed for the integration of the Franconian mining administration into the Prussian system, which the new Prussian minister Karl August Freiherr von Hardenberg (1750–1822) furthered through a reorganization of the mining administration in 1792.

On his inspection tour from July 12 to August 5, 1792 Humboldt visited mines, ironworks, peat bogs, vitriol works, the saltworks in Gerabronn, and the porcelain manufactory in Bruckberg. His observations were anything but encouraging. Franconian mining was in a state of "greatest decay," he wrote to his former teacher of mine surveying in Freiberg, Johann Friedrich Freiesleben. Many of the formerly profitable mines had been closed and were now dilapidated, and the mining authorities had hardly undertaken any activities to change this state or foster technological improvements. Only iron mining produced a good yield. But here, too, the galleries in most mines were not deep enough, or were derelict as well. Moreover, there were the "miserable timbering" of the galleries, poorly ventilated shafts, and extraction methods that were so antiquated they did not even use gunpowder.[6]

In his official report, *Report about the Conditions of Mining and Smelting Works in the Principalities of Bayreuth und Ansbach* (*Bericht über den Zustand des Bergbaus und Hütten-Wesens in den Fürstentümern Bayreuth und Ansbach*), written in September 1792 and submitted in April 1793, Humboldt adhered to the standards of administrative reports, which he had learned at the Freiberg Mining Academy. In the introduction, he stated that he had taken pains "to collect statistical and economic information about the management of all the mines und smelting works."[7] Statistics here meant the systematic collection of all kinds of factual information expressed, where possible, in quantitative terms. Humboldt also paid great attention to the technology and natural conditions of mining.

His report was divided into four parts and several supplements. First came the description of ore deposits, including their geognostic characteristics (nine printed pages). In this part the man of science was speaking. In his personal correspondence from 1792, Humboldt went into more detail concerning his scientific observations. The second part was concerned with financial topics. The longest, third part included detailed depictions of mines, smelting works, peat bogs, the porcelain manufactory at Bruckberg, and the vitriol and alum works at Crailsheim. Humboldt wrote this part in the chronological order of his tour based on notes he had taken, using the technical mining terminology of the time. Above and beyond this, his report also contained a summary of his observations along with assessments of the state of the mines. In this context, Humboldt also made suggestions about "means to a more flourishing operation of the mines."[8] His report thus combined the perspectives of a man of science, a mining expert, and an official who took into account state finances and the social organization of mining.

The Franconian mining administration was divided into three mining districts, and Humboldt elaborated on each of them separately. To get an idea of his evaluations, we first consider the Naila district, where iron and copper were extracted. "The mining operations here," Humboldt summarized, "require a complete reform."[9] The most pressing issue was the construction of a deeper gallery, to be funded by the royal treasury. He added suggestions about the location where the gallery should be constructed and about its length and depth—namely, 80 *Lachters* (approximately 160 meters). The proposal was designed to incur the lowest possible costs for the draining and timbering of the mine.[10]

The Wunsiedel district produced a better balance sheet, thanks not least to its rich iron deposits. Iron extraction there yielded, beyond the ten percent due to the shareholders, "a pure surplus of 900 thalers in the Royal Treasury." The royal earnings had also increased on account of the introduction of the *Bergseidel*, a standard unit of measurement for the quantification of ore. All in all, this district had an "active trade," with about a third of the revenues from iron mining generated by external trade, which was a presentable mercantilist achievement. Humboldt did not forget to mention that this achievement was partly owed to "the activity and the spirit of order" of the current district mining master, named Schubert.[11]

However, there were also some deficiencies in the Wunsiedel district. Humboldt criticized that all limestone quarries were leased. "At least a few should be operating on the royal account," he argued, for this would put money in the state treasury and furthermore allow greater savings in wood.[12] In addition, he suggested the introduction of wood-saving timbering of the mine shafts.[13] The shortage of wood was a permanent problem in the mining of the day, and Humboldt also highlighted the waste of wood in other passages of his report. Another suggestion for technical improvement concerned reducing the cost of transporting ore through changes in shaft construction and in the means of transport. In Freiberg the "dogcart" (*Hunt*) had been introduced for ore transport, a box-shaped, four-wheeled tram that ran on two wooden rails. By contrast, in the Wunsiedel district all ores were still transported with older types of carts, which implied a waste of manpower. Humboldt also pointed out that there were technical problems in the ventilation of the shafts and galleries—ventilation still consisted of waving fir branches—and in the methods of extracting ore with hammer and chisel, while it would have been possible to use gunpowder.

Humboldt added some criticism concerning the social organization of labor. The foremen did not have enough authority, and some of them collaborated "more closely with the hewers than with the mining administration." The latter smelt of fraud, which was a recurrent problem for the mining administration. Humboldt recommended another form of remuneration and a replacement of the current twelve-hour shift by the Freiberg system of an eight-hour shift. The twelve-hour shifts, he criticized, only "encouraged laziness." Humboldt had observed that the miners began their twelve-hour shift with a hearty breakfast before entering the mine, so that they "were hardly seven to eight hours on site."[14]

THE BUSTLING LIFE OF A MINING MASTER

The results of Humboldt's inspection tour must have made a strong impression on Heinitz and Hardenberg. In late August 1792, they offered Humboldt the position of chief mining master (*Oberbergmeister*) in the principalities of Ansbach and Bayreuth. Just before his twenty-third birthday, Humboldt thus was to take on the supervision of all of the principalities' mines, smelting works, and mine factories. Heinitz and Hardenberg negotiated that Humboldt was to remain in Franconia for only two years, during which he would continue to inspect Prussian saltworks for the Berlin administration as well. In addition, Humboldt was allowed to travel privately at his own expense. At the time it was not unusual for a leading official to be granted a leave of absence, not least because the authorities hoped to gain useful information from private travel as well.

Humboldt was overwhelmed by the offer. On August 27, 1792 he wrote to Carl Freiesleben (1774–1846), his closest friend and fellow student at the Freiberg Mining Academy: "All of my desires, good Freiesleben, are now fulfilled. I will now live entirely for practical mining and mineralogy ... I am reeling with joy."[15] An even more euphoric letter followed on the day of his formal appointment on September 6, 1792.[16] Humboldt was not out, as has been claimed countless times in the literature, to rid himself of his official post as quickly as possible, so that he could dedicate himself completely to pure science. On the contrary, he was eager to engage in practical mining, and he was determined to use his inspections for natural-scientific studies as well. During his time as a mining master, Humboldt would not miss a single opportunity, during his visits to the mines, inspection tours, and even his diplomatic missions, to make mineralogical and geognostic observations.

Nearly a year passed before Humboldt took up his official duties in the Franconian provinces. In the meantime, he was occupied with an eight-month inspection tour on Heinitz's behalf, which took him to Munich, Salzburg, Linz, and Vienna and from there to Silesia. The special goal of his tour was the inspection of saltworks, but the porcelain manufactories in Nymphenburg and Vienna were also on his agenda. Humboldt spent twelve days at the Reichenhall saltworks, where he received instruction from the facility's director, Johann Sebastian Claiß (1742–1809). Claiß helped Humboldt to complete a map of Germany's salt springs that he had already begun on his first inspection in Franconia. With this map Humboldt attempted to establish relationships between salt springs and rock formations. As before, Humboldt thus used this official tour also for natural inquiry.

In late May 1793, Humboldt moved to Franconia to start his bustling life as a mining master. Just a few days after commencing work in his new office, he wrote to Freiesleben: "I've just come back from the mine. I rode on horseback two miles and spent nearly three hours underground in the *Fürstenzeche* [a mine], so do not be surprised, dearest Freiesleben, if I write you a confused letter. Mining is improving faster than I thought. The preliminary organization is almost managed, the central mining office opened, the budget of the mining aid fund (*Bergbauhilfskasse*) established, and now I will concentrate on the individual mining offices."[17] One of Humboldt's first technical tasks was to restart mining at the *Fürstenzeche*, a mine administered by the mining office of the district Goldkronach. As the name suggests, this mining district had previously been rich in gold (and in silver), with a long tradition of mining gold and silver ores going back to the Middle Ages. By 1793, most parts of the *Fürstenzeche* were closed. In his inspection report of September 1792, Humboldt had described the state of this mine in detail and commented on the "prospects for the continuing process of reopening [*Wiedergewältigung*] the *Fürstenzeche*."[18] He reported that the mine was still passable, but its reopening, which included the removal of fallen rocks and floodwaters, had been given up a quarter of a year before "on orders from above." Humboldt had inspected several galleries and noticed that the desolate condition of the mine conflicted starkly with the official description by local mining authorities. He then recommended resuming its less expensive parts and pointed out that it "might be possible to realign the vein, which deserves great attention due to its gold and silver content."[19]

At the time, Humboldt did not imagine just how much effort gold mining would cost him in the following years. Although Heinitz was skeptical about this undertaking, he finally approved test mining, with support by nine miners. Hardenberg, who fully supported Humboldt's project, also authorized him to carry out technological experiments on gold extraction using mercury. In winter of 1793, Humboldt thus began his first experiments on the amalgamation of gold ores. Shortly thereafter he reported to Freiesleben, "after minor experiments they [the gold ores from Goldkronach] amalgamate well."[20]

During his stay at the Freiberg Mining Academy, Humboldt had visited the amalgamation plant in Halsbrücke, run by the Mining Academy's professor of mathematics Johann Friedrich Wilhelm von Charpentier. There a new method was being used to extract silver from its ores, the so-called cold amalgamation process, which had been developed by the chemist and Hapsburg mining official Ignaz von Born (1742–1791).[21] Many chemists and mining officials regarded this method as a proof of the practical utility of chemistry. As we have seen, in the late 1780s, Minister von Heinitz had sponsored amalgamation experiments by the mineralogist and mining official Johann Jacob Ferber in the newly established laboratory of the Berlin mining administration. When Ferber died unexpectedly in 1790, these experiments ended abruptly without having yielded clear results. This failure may have caused Heinitz to be cautious when Humboldt requested his support three years later. Even so, Humboldt continued his amalgamation experiments in the subsequent years. In winter of 1796–1797, at the end of his term in Prussian state service, Humboldt and Hardenberg were optimistic that they were close to a breakthrough for the introduction of a new amalgamation method for gold, although the question of whether it would actually be economical on a large technical scale had not been settled conclusively at that time.

The year 1793 also brought scientific success. In June 1793 Humboldt was admitted to the Leopoldine Carolinian Academy, and in September of the same year he received a diploma of admission to the Gesellschaft Naturforschender Freunde zu Berlin in recognition of his "laudable zeal for natural studies [Naturkunde]."[22] Time and time again the young mining master found opportunities for mineralogical and geognostic observations, which he reported to his scientist friends. For example, in July 1793 he assured Freiesleben that he would certainly "not forget geognosy." On the contrary, on his daily visits to the mines he often observed unusual minerals and rock formations. In early August he informed his Berlin friend Dietrich Ludwig G. Karsten that his daily observations enabled him to recognize geognistic patterns. "My ideas expand more and more," he wrote, "I begin to draw from nature, I see similarities in the layers [of rocks] and in their orientation." A few months later, he decided to write "something big geognostic," based on his observations.[23]

In the spring of the following year, Humboldt set off on an inspection tour, visiting saltworks on behalf of the Berlin mining administration, which took him to South Prussia and Upper Silesia. In July, he had to accompany Hardenberg to the headquarters of the Prussian army in Frankfurt am Main. After the defeat of the Austrian-Prussian coalition's antirevolutionary offensive in France (at Valmy) in September 1792, the French army had occupied the left bank of the Rhine with the cities Speyer, Worms, Mainz, and Frankfurt am Main. In December 1792, the Hessian and Prussian armies had reclaimed Frankfurt, and the Prussian army remained there in position while negotiations were being conducted. Hardenberg played a key diplomatic role in these negotiations. In July 1794, he proceeded to the Rhine, and Humboldt, whether he liked it or not, had to accompany him. "The businesses of the army disturb me greatly," Humboldt wrote to Freiesleben, and to his poet friend Friedrich Schiller he complained bitterly about the "restless travels" with the minister.[24]

But the diplomatic mission also offered a good opportunity to inspect mines in the vicinity of Frankfurt and to deepen his geognostic knowledge. Thus, he wrote to Freiesleben: "I gained some new ideas, and the constant travels in mineralogically interesting areas helped me greatly in writing my book on layers and orientations [of rocks]."[25] The geognostic project Humboldt had planned in the previous winter became more concrete. After his return to the army camp in Frankfurt, he sent a chapter of his book manuscript to Freiesleben, which also enclosed geognostic drawings, and asked him for comments.[26] Only some three decades later did Humboldt actually publish a work on the stratification of rocks, his *Essai géognostique sur le gisement des roches dans les deux hémisphères* (1823).

In November 1794, Humboldt received an offer from Heinitz to become the chief mining master in Silesia, but he declined.[27] After a petition for dismissal in March 1795, he received another letter from Heinitz in April 1795 in which the minister tried to convince him to stay in Prussian service. In this letter, Heinitz appealed to Humboldt's often-repeated desire to serve his fatherland and to do practical work in mining, emphasizing that the practical service of an official and the sciences mutually complemented each other. If you dedicate yourself to mining, he wrote to Humboldt, "you can and will make yourself useful to the sciences at the same time [and] accumulate experiences for them; or [you can] confirm observations or refute proposed theories, in some cases by yourself, in other cases by encouraging others, and thus achieve the ideal of the man serving the public benefit [*das*

Ideal des gemeinnützigen Mannes]."[28] Appealing to the ideal of the man serving the public benefit, Heinitz highlighted the credo of a new generation of Prussian officials: that the sciences, in promoting technical progress, also improved the common good.

Shortly thereafter Humboldt consented to remain in Prussian service. From July to November 1795 he received a leave of absence for a long-planned trip to the Alps. On the scientific agenda of the journey were botany, mineralogy, geognosy, and, for the first time, also the physics of the earth. But Humboldt did not conceive of this journey as a purely private undertaking. On the contrary, he sent a travel report to Heinitz, in which he described among other things innovations in saltworks.[29]

STUDIES OF GASES AND TECHNICAL INVENTIONS

In November 1795 Humboldt began to work systematically on two technical inventions: a miners' lamp and a rescue device or "respirator" for miners.[30] This inventive work is a prime example of his shifts from natural to technological studies and vice versa. During his time at the Freiberg Mining Academy Humboldt had already studied the airs in mines, or in the miners' language mine "weather," and measured their temperature. From the beginning of his employment as a mining master he continued these studies and extended them to the measurement of the mine airs' humidity and their electrical and magnetic charges.

As Humboldt and his coexperimenters were familiar with the suffocating effects of mine airs, they often used oxygen bottles during their experiments. It was in this context that Humboldt got the idea to construct a respiration machine and a new miners' lamp that worked with oxygen.[31] In July 1795, his research on mine airs led to a first publication in the journal *Chemische Annalen*. In this essay, Humboldt argued for the foundation of a new discipline that he called "subterranean meteorology," which would comprise the study of mine airs along with the winds, humidity, lightning, and other weather phenomena they produced in the mines. "Nature does not know above and below ground," Humboldt emphasized in his publication.[32]

After an interruption due to his private travels, Humboldt resumed his studies of mine airs in November 1795, now also embarking on his work of invention. He wrote about the relationship between his subterranean meteorology and his work of invention: "I was not content with a study that would expand our physical knowledge but be useless for practical mining." Therefore, his "eager desire" was "to invent means that reduce the disadvantage [that bad mine airs have] for human life and the business of mining."[33]

Humboldt's miners' lamp consisted of an oil lamp, an oxygen container, and an adjustable air tube connecting the two parts. The lamp burned in oxygen-poor mine air, in which the normal miners' lamps went out, and it was supposed to be used in all mines. The respirator, by contrast, was a rescue apparatus that was to be used only in special cases. It consisted of an oxygen container, a tube, a mouthpiece with valves that separated inhalation and exhalation, and a breathing mask, and was to make rescue efforts possible in poisonous air mixtures after a mine accident (see figure 10.2). In addition, Humboldt intended it for military use, specifically by the *corps de mineurs*, whose task was to dig tunnels under the fortifications of a besieged fortress and plant gunpowder mines in them.

In the spring of 1796 Humboldt wrote to Freiesleben: "The invention of my respiration machine is largely completed. I have valves that are opened by just a touch of air, and one can breathe with it in any air for three hours. The air sac is filled with common air. The [miners'] lamp burns even in fixed air [in later terms, carbon dioxide]." For practical reasons, Humboldt had substituted common air for the oxygen of his respiration machine, although this change required a larger air container. Concerning the uses of his machine, he added: "The General War Department [*Ober-Kriegs-Collegium*] investigated the thing and it passed."[34] Humboldt had involved the General War Department in the practical testing of his respiration machine, and he wanted to carry out additional experiments in the mines of Goldkronach before he made his inventions public.

In mid-July 1796, Humboldt had to accept another longer interruption of his work, since he once more was called to diplomatic service in the war against the French. Not until September was he able to continue his inventions. The respirator and miner's lamp now entered a systematic practical test phase in mines. Among other things, Humboldt tested the substitution of common air for pure oxygen in his miner's lamp. From the very beginning he had invited mining officials and miners to participate in his experiments. His closest partners were Eberhard Friedrich J. Killinger, who had studied at the Freiberg Mining Academy in 1791 and had just advanced to become a deputy mining master (*Vize-Oberbergmeister*), and the juror (*Berggeschworener*) Heinrich Ludwig Sievert. In mid-October, his partners prevented the deadly outcome of an accident. "Yesterday I nearly became a victim of my own experiments," Humboldt reported to Freiesleben. He had intended to test his lamp again in an alum mine, where it had failed

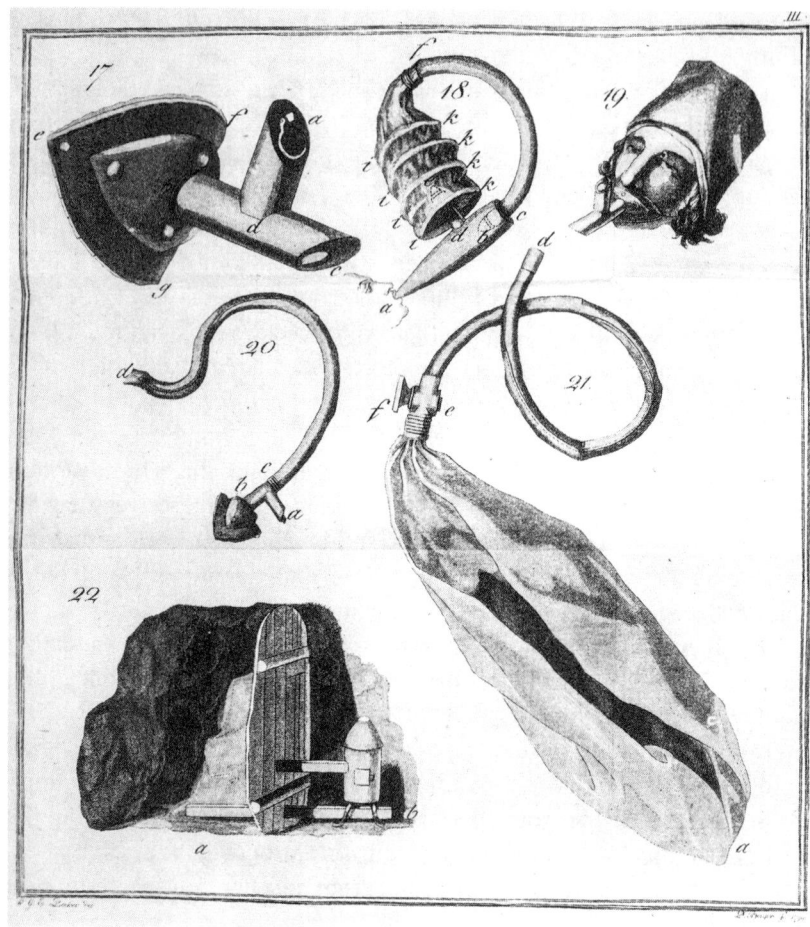

FIGURE 10.2
Humboldt's respiration machine.
Source: Humboldt 1799, plate III.

previously. At first everything went well, the lamp "burned brightly in the bad airs," but then things turned out differently:

> I was curious, wanted to go all the way up to the rotten wood where we had burned the sulfur. I crawled inside. Killinger had to remain behind, because he is still ill from a similar experiment he performed in the Naila district. I arrived at the location, set my lamp down and its light gave me immense joy. I became very tired, yes even reeling, I sank to my knees next to the lamp. I supposedly called Killinger, I know nothing of it. He groped after me in the dark and found me unconscious next to the lamp. He pulled me out.[35]

Humboldt also informed his friend that his group had achieved a break-through in the practical testing of the lamp:

> Since the day before yesterday we—I, Killinger, Sievert, all foremen and miners who were present for the experiments in the three districts—were convinced that my lamp is fully perfected for practical use and that it is time to send and assemble the large lamps. You cannot believe how I am reeling with joy about this. How many expensive apparatus will be avoided in this way.[36]

Shortly afterward, Humboldt published technical details of his inventions in the *Annalen der Chemie*.[37]

Parallel to his work of invention, Humboldt began to perform chemical analyses of gases. In mid-September 1796 he wrote to Freiesleben that he was experimenting in mines with pure gases and mine airs.[38] These experiments served two goals. First, Humboldt wanted to expand his subterranean meteorology to include knowledge about subterranean combustion and about the chemical composition of different kinds of mine air. Second, knowledge about the composition of mine airs was crucial for testing the functionality of his lamp. During his tests of the lamp in the alum mine, mentioned in the letter quoted above, the lamp extinguished repeatedly. This could have been for two different reasons: either the experimenter's mechanical handling of the lamp was wrong, or the oxygen content of the mine air was extremely low and the lamp's air supply needed to be improved. The analysis of the mine air with a eudiometer proved that the latter was the case. Hence, Humboldt also reported to Freiesleben about his idea for "how more air can be transported to the flame" of the lamp.[39] A new device was needed that delivered more air to the flame without extinguishing it simultaneously. The solution to the problem consisted of the construction of an annular tube with numerous small openings, which replaced the simple opening of the main tube connected with the air container of the lamp.

The chemical analysis of different kinds of "air" or "gases" was a flourishing field in late eighteenth-century chemistry that had been established only in the middle of the century and comprised both conceptual and technical innovations. "Air" had long been considered as a simple element, and chemists had thus assumed that there was one kind of elemental air. The new field of "pneumatic chemistry" also brought a plethora of new instruments and techniques into the chemical laboratory. It required familiarity with the experimental methods of preparing and purifying different kinds of air (gases), and technical know-how about the construction and manipulation of related instruments and apparatus. Furthermore, the analysis of gases involved chemical knowledge about the interaction between gases and reagents, qualitative methods for the identification of different kinds of gases, and quantitative methods for the determination of the proportions of the different components in a gaseous compound or in a mixture of different gases. The most important instrument allowing the latter was the eudiometer.

The English chemist Joseph Priestley had introduced the eudiometer to measure the "vital air" (oxygen) contained in common air. In September and October 1796 Humboldt tested different types of eudiometers and built replicas of the eudiometers by Heinrich Reboul and Guyton de Morveau, which used solid reagents for binding oxygen. De Morveau's eudiometer, he reported to Freiesleben, "is easy to make oneself and magnificent for investigating mine air!" Shortly afterward, he wrote, "I am going down [into the mines] diligently and now am really conducting experiments about the analysis of the weather [mine airs], which are quite striking. For I have simplified Reboul's phosphorus eudiometer such that I can make the instrument myself." Performing chemical analyses of mine airs, Humboldt transformed the mine into a chemical laboratory. He had constructed a small, portable eudiometer, which was easy to carry on horseback and robust enough to survive difficult descents by ladder into a mine. With unmistakable pride he related to Freiesleben the advantages of his instrument. "This is my setup. Can there be anything safer and simpler?" He immediately added that he was making "great plans for a book on eudiometry."[40] The book should contain measurements both below and above ground. Humboldt had already started to carry out eudiometric measurements of the air on the surrounding mountains. For the future he planned to build measuring stations over wide areas. First of all, however, he recommended that his friend build a replica of the instrument and conduct his own experiments in the mines of Freiberg, not least in order to check his own subterranean experiments.

By the end of 1796, Humboldt was fully convinced that his invention was completed. He had carefully prepared and planned his chemical and technological experiments, and he took notes on the experiments, leaving a comprehensive manuscript. In 1799, Wilhelm von Humboldt published his brother's manuscript in book format. In the introduction, he pointed out that his brother's "position as chief mining master gave him the opportunity to amass a greater number of interesting observations and to distinguish the feasible from the infeasible more surely and better than mere theorists would have been able to."[41] In one sentence Wilhelm von

Humboldt thus highlighted two characteristics of his brother's life as a mining master: his position made it possible for him to accumulate a number of interesting natural observations, and to acquire practical technical skills and judgment about the technically possible.

MINING EXPERT AND MAN OF SCIENCE

In February 1797, Humboldt quit state service. In the literature on Humboldt it has been claimed unwaveringly that his activities as a mining official were an obstacle to his studies of nature, and that this was the reason why he quit state service in February 1797, when, following the death of his mother in November 1796, he got the financial means to do so.[42] Yet Humboldt's own statements, including those in his most private letters to his friend Freiesleben, do not support this view. The reasons why he gave up his position in spring of 1797 are of a more complex nature, and technical work was not among them. Humboldt often complained about intrigues in the state bureaucracy, its hierarchic organization, and its rules of strict obedience and subordination that were enforced not least by Heinitz. By contrast, he pointed out repeatedly that his mineralogical, geognostic, botanical, and chemical inquiries were facilitated by his activities as a mining master, even though his position ruled out longer travels overseas.[43]

One of Humboldt's final official acts was to complete a report about the condition of mining in Franconia, entitled *Brief Description of the Current Conditions of Mining in the Franconian Principalities* (*Kurze Darstellung der gegenwärtigen Verhältnisse des Bergbaus in den Fränkischen Fürstenthümern*). In this report of February 24, 1797, Humboldt took stock of the improvements he had achieved. Its general tone was inspired by the ideal of the man serving the public benefit. Franconian mining was very important "for the welfare and industriousness of the people," Humboldt wrote, but it was still backward, although the mining folk were loyal, patriotic, and cheerful. The state, he further pointed out, had to give up the goal of fast money making and monetary accumulation, as the old cameralists had intended, and instead search for means to achieve sustained economic improvement. In this context he observed that it had been a general mistake of past mining in Franconia not to build mines deep enough and to extract only rich ores, "which makes future extraction of the iron ore in greater depth partly more difficult, partly impossible."[44] But Humboldt was also able to report some improvements. In the Goldkronach mining district the extraction of gold had considerably increased, and after numerous trials it could be hoped that amalgamation on a large technical scale would be introduced soon. In the Wunsiedel district new wood-sparing timbering methods had been introduced, an old gallery had been restored, and the construction of water wheels had begun. As for the Naila district, the mining administration had begun to construct a new, deep gallery, named the "Friedrich-Wilhelm-Stollen" after the Prussian king. Humboldt had elaborated the construction plans himself and "calculated everything down to the nails [*Spindenägel*]."[45] The Friedrich Wilhelm Mine had the potential of "opening up the entire district, and securing mining around Steben, which is so intimately connected with the prosperity of the countrymen, far into the distant future."[46] Here again, what Humboldt highlighted was the public utility of state-directed mining.

In his report Humboldt also emphasized the need for scientifically educated mining officials. "All means must be mobilized," he wrote, "to recruit skilled and industrious foremen, jurors, and mining masters."[47] Humboldt had succeeded in gaining royal support for two aspiring mining officials to study at the Freiberg Mining Academy. With unmistakable pride he also reported about the mining school for "the common mining people" he had set up in late 1793 in Steben, the village where he lived. In the German mining regions, it was customary for all boys aged eleven to twelve to start working in the mine, where they initially performed the physically less demanding jobs like sorting ores. As we have seen in the previous chapter, Heinitz organized the foundation of local mining schools in Prussia to prepare the boys for their occupations through additional schooling. Humboldt emphasized that his own mining school, which complemented the village school, trod new paths. Its purpose was not to drill the pupils in the hard work of mining, but to "combat superstition" along with the propagation of "reasonable concepts," so that the miners would not "undermine their prosperity through foolish undertakings."[48] Instruction included writing; basic mathematics and practical mathematics, such as the use of the compass, measuring, surveying; the basics of geognosy; mining law; and the history of mining in Franconia. Humboldt also put together a small mineralogical cabinet and wrote a textbook to support the teacher, the young foreman Georg Heinrich. The school, which became known as the first "Royal Mining School" in Franconia, was apparently a success. In 1797, thirty to forty pupils attended lessons. There were two classes, one for the young mining boys, who came in the afternoon, and another for the somewhat older miners, whose instruction took place in the evenings. As Humboldt proudly pronounced in his report, several good foremen had already been trained in his school.

In the five years of his term as a mining official, Humboldt was a mining expert, inventor, organizer of a mining school, and man of science. In his eyes, mines were fascinating technical objects as well as locations for observing nature and performing experiments. Observing layers of rocks, collecting minerals, and carrying out experiments in mines, Humboldt became a man of science known beyond the borders of Prussia. Hardenberg later observed that Humboldt, "by going down into the mines, at the same time studied the chambers of the earth."[49]

III USEFUL SCIENCE
AND ITS PRACTITIONERS

11 MINING SCIENCE

In the previous chapters, we have studied practices of mining science, especially experimental research and work on inventions. We will now turn to mining science as a body of knowledge and a teaching discipline and ask, what were its epistemic components and overall aims? Further, what did "science" mean in this context? Textbooks can provide a first overview of these issues. In his *De re metallica* (1556), Georg Agricola already refers to a "*Wissenschaft vom Bergbau*"—a science of mining. The suffix *-schaft* here indicates the comprehensive scope of the body of knowledge (*Wissen*) relevant to mining.[1] In the eighteenth century, Agricola's book was still used for teaching, but new textbooks began to replace it. These included Friedrich Wilhelm von Oppel's *Bericht vom Bergbau* (1769), a textbook used at the Freiberg Mining Academy, and Christoph Traugott Delius's *Anleitung zu der Bergbaukunst* (1773), used at the Mining Academy of Schemnitz.[2] Oppel's textbook begins with the following definition: "Under the art of mining [*Bergbaukunst*] we understand that science [*Wissenschaft*] which teaches us the most advantageous extraction of the fossils present in our earth."[3] The definition shows that the historical actors did not differentiate between "art" and "science" in this context and that alongside the term *Bergwerkswissenschaft* (mining science)—and *Bergwissenschaft* and *bergmännische Wissenschaft*—the terms *Bergbaukunde* and *Bergbaukunst* were also used for designating the new science of mining.

Oppel's and Delius's textbooks concentrate, for specific didactical purposes, on mine construction and other parts of mining technology, while metal smelting and processing as well as large parts of the natural sciences and mathematics are left out. Only in the first chapter of his book does Oppel deal with geognosy and mineralogy. Likewise, Delius refers in his introduction to the wide scope of mining science, which, he points out, comprised mine construction, machinery, and mining techniques as well as "geometry, trigonometry, aerometry, mechanics, hydraulics, and hydrostatics," and "mineralogy and metallurgical chemistry with application of their principles to the art of assaying [*Probierkunst*] and to the whole of smelting and mining manufacturing."[4] However, orienting himself toward Oppel, the bulk of his book is concerned with mining technology in the narrow sense and with mining economics (*Bergbauwirtschaft*). A more comprehensive textbook is Johann Thaddäus Anton Peithner's *First Principles of the Mining Sciences* (*Erste Gründe der Bergwerkswissenschaften*), published in 1769–1770. Peithner, who held the chair of mining science at the University of Prague, aimed to present "the foundations of all mining sciences pragmatically and in a systematic order." His textbook comprises mining and smelting technologies, including the arts of surveying and assaying, as well as natural sciences such as subterranean physics (*unterirdische Naturlehre, physica subterranea*), subterranean geography, mineralogy, and metallurgical chemistry.[5] Conceived as a rudimentary manual for beginners, it consists of a series of short paragraphs without any annotation; thus, it provides only limited insight. The following analyses are based on different sources that were not only unintended for publication but also reveal a more detailed picture of the historical actors' understanding of mining science.

HEINITZ'S UNDERSTANDING OF MINING SCIENCE

In 1771, six years after the foundation of the Freiberg Mining Academy, Saxony's General Mining Commissioner and cofounder of the Mining Academy, Friedrich Anton von Heinitz, wrote a long audit report in which he also addressed the aims and content of the science of mining. Having praised the role played by the state in Saxon mining, he pointed out that technical improvement was possible only with systematic planning and the organization of mining science. With respect to the state of mining in Saxony, he observed that the following three components of mining science were the most important: the drawing of mine plans; studies of the earth; and mineralogy and metallurgical chemistry in connection with ore assaying and metal smelting.[6]

Heinitz viewed the drawing of "reliable mine plans" as an indispensable task of the Saxon mining authorities. He thus pointed out that the local mining offices would have to take "continual care" for the improvement of the mine surveyors' instruments, measurement techniques, and drafting skills. Based on his inspections, he noted caustically that in some offices mine plans did not exist at all or had been drawn up only very recently, after his admonishment.[7] The second component of mining science—studies of the earth—ranged from geognosy to physical geography and geological mapmaking. With his call for a countrywide investigation of the "rock masses and mining regions" and the geological mapping of Saxony, Heinitz adopted a point of view already expressed by Agricola in *De re metallica*, but one that apparently had been ignored by the Saxon mining authorities.[8] Hence, in the 1740s Zimmermann had been still calling for "a complete montane geography," including geognosy as well as the topography of the landscape and even the climatic conditions. Such an endeavor encompassed the comprehensive description of the mining regions and whole "regions of the country" and subsequent mapping thereof: "A miner also must have a complete understanding of the whole region of his mining operations," Zimmermann had observed, "and beforehand have a probable idea of where to start building a good structure." To get a clear idea of the region, he had added, "one can never have enough general and specialized maps."[9] In his report, Heinitz reemphasized this argument. As a result, still

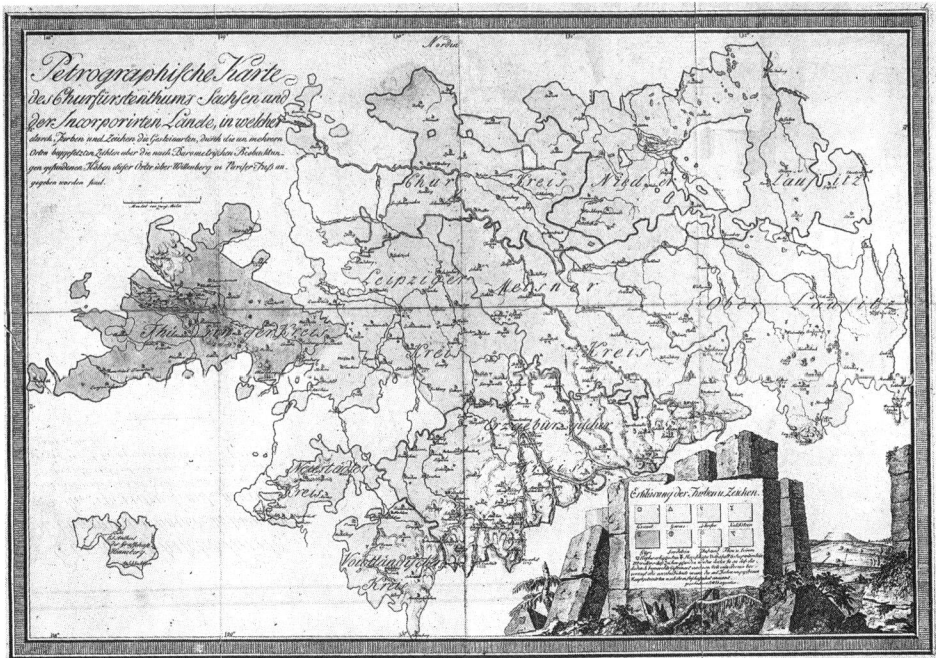

FIGURE 11.1

Charpentier's "petrographical map" (1778).

Source: Sächsische Landesbibliothek—Staats- und Universitätsbibliothek Dresden, no. df_Id_0020994.

in the same year (1771), the mining authority ordered the mining councilor and professor at the Freiberg Mining Academy, Johann Friedrich Wilhelm von Charpentier, to prepare "a mineralogical map of the land of the
elector of Saxony." The map was completed in 1778 (see figure 11.1).[10]

As to geognosy, Heinitz pointed out that it should study the precise
"investigation and knowledge of the rock masses [Gebirge]," including their
composition, spatial structure, age, and "causes of origin." Heinitz observed
that the oldest rock masses consisted mainly of granite that contained little
or no silver ore; thus he created a correlation between the age and kind of
rock mass and the occurrence of silver. Granite was overlaid by the ore-rich
Mittelgebirge, composed of gneiss and slate, which in turn were overlaid and
broken up by ore-bearing seams. The cause of the formation of these younger
types of rock masses, Heinitz added, lay in the "individual inundations and
floods, and perhaps also fire-spewing mountains."[11] Geognosy thus yielded
knowledge about the natural conditions of mining, especially about the
relationship between the age and kind of rock masses and the kind of ore
deposits. Based on knowledge about the structure of rocks, it also provided
information about the location of ore deposits and thus for prospecting.

For Heinitz, the issue of prospecting for new ore deposits was particularly
important, for he saw an acute lack of knowledge in this part of mining.
He was a vehement opponent of the dowsing rod, which he regarded as a
striking example of untrustworthy traditional practices. By contrast, geognosy opened up new, more reliable ways of prospecting. Once the spatial
possibilities of ore deposits had been narrowed down by means of geognostic knowledge, it was more likely that surface explorations would find the
"exact point" of the ore deposit. Prospecting based on geognosy, Heinitz
emphasized, followed the "order of nature."[12] The "order of nature"—a key
concept of eighteenth-century natural history—was reflected accordingly
in the large-scale distribution of ore deposits and their integration in characteristic rock types and rock formations. This kind of natural knowledge,
which would lead to a complete identification of all the ore deposits in a
country, was relevant for at least two groups of mining officials: first, the
mining masters and councilors, who organized prospecting; and second,
the mine surveyors, who needed to be familiar with the mine region and
provided information on the ore deposits in their mine plans.

The third cornerstone of mining science was knowledge about the properties and behavior of raw minerals and processed metals, which was provided by mineralogy and chemistry. Mineralogy and chemistry were stores of
knowledge about materials, which flowed into the work of leading officials,
who made decisions about the mining of new metals and improvements

in metal smelting as well as the work of the assayers and smelters. In his report, Heinitz pointed out that Saxony now mined the silver-poorer "barren" ores, and that the finding of a new method for the extraction of silver from such ores was a technical challenge that he hoped to master with the "recent felicitous progress in mineralogical and metallurgical-chemical knowledge."[13] Metallurgical chemistry was a chemical subdiscipline that was still in its infancy at the time, and as a practical man Heinitz certainly did not assume that it would offer rules and formulas that merely had to be applied in silver smelting. He was also familiar with the fact that metallurgical smelting experiments first had to be scaled up to become practically useful. Hence, his argument on the usefulness of metallurgical chemistry was more nuanced. The metallurgical chemistry of his time had, he observed, "attained a much higher degree of clarity and certainty [*Deutlichkeit und Gewißheit*]" than ever before, and one ought "not fall behind" this development.[14] Formulations like these are evidence of Heinitz's reflection on the relationship between theory and practice.

Mining science, Heinitz emphasized, did not produce absolutely certain knowledge, but rather knowledge with a "higher degree" of certainty than the locally embedded practical knowledge of the miners. The mining authorities had to advance this kind of science, since it was the most advanced knowledge and the best tool to foster improvement of mining. Hence, Heinitz expected from metallurgical chemistry not a scientific revolution in metal smelting, but rather more reliable knowledge that would help to achieve gradual technical improvements. This understanding of "science" differed profoundly from the epistemology in the framework of eighteenth-century natural philosophy. The hidden principles and causes of natural phenomena that natural philosophers aimed to lay bare were not considered to be just relatively certain and provisional truths. On the contrary, the ultimate goal of natural philosophy was absolute certainty of knowledge codified in the language of logic and mathematics, and absolutely true natural laws as well as theories about the ultimate structure of matter.

Heinitz did not spell out in his report what exactly he meant by reliable knowledge, but he gave some hints.[15] For example, he pointed out that "more careful calculations in all smelteries" were needed, as well as systematic comparisons of the metal yields at domestic and foreign smelting works. Another issue he highlighted was the controlling of the smelting process through repeated and precise chemical assays and accompanying calculations of the metal yield. Furthermore, he exhorted the mining officials to "industrious experimentation and the keeping of a journal." The journal served to protocol the experimental data, including the "notation of success" and "probable causes." Reliability of knowledge thus hinged not least on the acquisition of scientific methods—precise quantitative methods, calculation, repeated control of processes, experimental testing, and the writing down of experimental results. The method of quantitative chemical analysis, which at that time was making its entrance into mineralogy, and the keeping of written protocols served as models for the way of acquiring reliable knowledge; in the late 1780s, Heinitz would also introduce these methods into the Royal Prussian Porcelain Manufactory (see chapter 4). Further, written reports were an excellent tool for the communication of experimental results among the corps of mining officials, which would in turn facilitate intersubjective control. If one combined scientific methods with a system "of extraordinary rewards," Heinitz added, one would be on the path to the systematic organization of inventions and improvements of mining.[16]

Unlike the goal of true and certain knowledge in the framework of natural philosophy, reliable knowledge was not a goal in itself but rather a tool for other, practical goals. Thus, reliability of knowledge is not merely a more modest epistemological goal than truth and certainty, it also has different connotations. The traditional scholar strove for certainty of knowledge, not least since knowledge represented the inner spiritual nature of man. Certain knowledge was the highest form of knowledge and the most perfect achievement of the human mind. Hence, for the traditional scholar it made no sense to ask questions about the function of certain knowledge. By contrast, "reliable knowledge" belongs to a pragmatic understanding of knowledge, implying that knowledge is not acquired for its own sake but has a function in doing and making. Comparable to tools and techniques, reliable knowledge contributes to successful action. In line with this, "reliable knowledge" does not imply an expression of deep truths. Instead of truths, pragmatists like Heinitz wanted to get things right, which required circumscribing manageable areas of inquiry as well as measuring and analyzing properties and features of distinct objects and their functional relationships.

GERHARD'S CURRICULUM FOR MINING SCIENCE

The considerations on mining science that Heinitz discussed in his audit report from 1771 touched merely on those aspects that he held to have priority in Saxony in the short and mid term. They were neither a comprehensive description of all components of mining science nor a detailed curriculum for the Freiberg Mining Academy. This section examines Carl

Abraham Gerhard's plan for the teaching of mining science in Berlin. As we have seen in chapter 9, this plan never became reality. Even so, our analysis of Gerhard's plan further illuminates the meaning of the science of mining.

In January 1770, on the order of Minister vom Hagen, Carl Abraham Gerhard drafted a plan for the establishment of a "complete school of mines" in Berlin, partly following the Freiberg model but also deviating from it. The plan included, alongside organizational proposals, a compilation of the knowledge components of mining science, as well as arguments as to why they were useful for mining officials.[17] The Freiberg Mining Academy, cofounded by Heinitz, had been one of the first establishments in Europe to institutionalize the science of mining. The founding of such an institution raised new epistemological and pedagogical questions. What type of knowledge and know-how constituted the mining officials' expertise and inventiveness? How broad and comprehensive should their education be, and in how much depth should they go in particular fields? How could the teaching of theory be combined with practice? There was a broad agreement at Freiberg that a balance needed to be maintained between theory and practice, that is, between knowledge that could be taught in the classroom and practical training in mining. Moreover, classroom teaching of technological and scientific knowledge had to include experimental demonstrations, demonstrations in the mineralogical cabinet, and work with models.

In contrast to the Freiberg Mining Academy, where classroom teaching was combined with daily practical training in mines, Gerhard proposed a division of the training of Prussian mining officials into two phases. In the first, two-and-a-half-year-long phase, all candidates would learn their profession "systematically and scientifically," that is, through classroom teaching, at one central location. His plan of January 23, 1770, thus stated that a mining official ought to "first and foremost be equipped with the necessary theoretical knowledge before he begins to beneficially practice his métier."[18] The Prussian capital Berlin seemed to be the right location for the theoretical part of the training, which would encompass not only lectures but also experiments, excursions, and exercises with models. The second phase would consist of a one-year practical course of training in the local mining districts.

Gerhard's proposals were tailored to Prussia, a country with mining regions that were widely dispersed and far away from the capital. The mining academy in Berlin would provide a common foundation of knowledge, whereas the acquisition of practical knowledge would be adapted to the particular requirements of local mining districts and professional responsibilities of mining officials. Over a decade later, Gerhard again took the initiative for the founding of a mining academy in Berlin, reiterating his proposal from 1770 to provide all the mining officials with the same scientific basic education.[19] This scientific education was to commence as early as possible, at the latest after a half-year-long course of practical training, since, as Gerhard pointed out, it was "a general experience ... that, if the soul of man once receives a practical pleasure [*plaisir*], it will be quite difficult to find theoretical knowledge to his taste."[20] Clearly, Gerhard's demarcation between practical "*plaisir*" and theoretical "knowledge" was not completely free of arrogance. What was also at stake here, however, was the tension between general education and specialized professional training. The same problem also was faced in postrevolutionary France, where the state officials first received a broad scientific education at the École Polytechnique and only afterward went for professional training to one of the *Écoles d'application*. And it still gives occasion for discussion today.[21] Creativity requires the capacity to look beyond the tip of one's own nose, but it also requires the closest familiarity with the methods and instruments of a specific discipline. The conflicting views on the teaching of mining science shows how old these problems are.

As we have seen in chapter 9, Heinitz's understanding of mining science emphasized the need for close temporal coordination of classroom teaching and practical technical training, as was practiced at the Freiberg Mining Academy, where morning visits to the mines were followed in the afternoon by "theory." In similar vein, at the Schemnitz Mining Academy, theory and practice were distributed over different days in the week.[22] Where such temporally close coordination was not possible, as in Berlin, Heinitz gave pedagogical priority to practice and deferred "theory" to a later phase of the training. Hence, he refused to support Gerhard's initiatives to establish a central Prussian mining academy in Berlin; instead, he ordered Gerhard to organize the foundation of local mining schools in the different mining districts.

Which knowledge components did Gerhard regard as belonging to the shared knowledge base acquired though classroom teaching? He summarized his credo thus: "The theory of mining and smelting is based on the knowledge of mathematics, mineralogy, physics, and the inextricably connected mechanics and hydraulics; further, on chemistry and in particular metallurgy; and finally, on the general knowledge of mine construction and smelting facilities."[23] Mining science thus comprised parts of mathematics and natural sciences as well as "general," or translocal, technological knowledge of mining and metal smelting. With respect to mining technology, Gerhard emphasized knowledge of the construction of mining structures, the building of shafts and tunnels, as well as the methods of excavation, transport, and ore processing. The basic technological knowledge of smelting included the construction principles of smelting furnaces, knowledge of fluxes and types of fuel, and knowledge of smelting techniques. Thus, mining science connected different knowledge components: technological knowledge, which at the time was compiled and generalized from technical knowledge embedded in local practices, and mathematical and natural scientific knowledge. In addition, Gerhard's plan included "mining economics" (*Bergökonomie*).

WHY NATURAL-SCIENTIFIC KNOWLEDGE?

What arguments did Gerhard present for the usefulness of mathematical and natural-scientific knowledge? Concerning mineralogy and the methods of ore identification, he stated: "Every miner and smelter needs to know the ores and minerals that he wishes to seek, extract, and process." To this plausible argument he added the following clarification. Because there were always "particular mixtures of the ores" in different places, without general mineralogical knowledge one would lack the orientation necessary to identify deposits in unfamiliar locations.[24] What could mineralogy contribute here? How could this lack of orientation be prevented? Here we should keep in mind that this argument did not apply to the simple miners but rather to the mining officials, who had to inspect many mines, organize prospecting for new ore deposits, and make decisions on whether ores were worth extracting. In Gerhard's argument, the mining officials who prospected for new deposits needed to have an overview of the possible varieties of a type of mineral in order to reliably identify a local variety. Such generalized empirical knowledge was provided by mineralogy and, in particular, by mineralogical collections. If one knew the characteristic properties and chemical composition of, for example, copper ore, and had already become familiar with several varieties of this kind of mineral in mineralogical collections, one would be well prepared for the identification of a special local variety in a new local deposit. Whereas if one only had a familiarity with one or two local varieties on the basis of traditional artisanal training, one would more easily err in the identification of a new variety. The practical-technical knowledge of the miner was local and depended on local conditions. By contrast, mineralogy, and natural history as a whole, tried to overcome the limitations of local technical knowledge by establishing unambiguous criteria for the identification of a mineralogical "species" and providing knowledge about its local "varieties."

Using similar arguments, Gerhard also made the case for the early acquisition of geognostic knowledge of the rock masses (*Gebirge*) and of chemical knowledge. The purely practical training of the miner was shaped by the contingent local conditions in the particular mines and smelting works he was introduced to. This entailed that his "knowledge is based solely on *isolated experiences* [*einzelne Erfahrungen*], which in every mining region [*Gebürge*] and at every smelting works are subject to *manifold changes*, which in turn require new experiences to be discovered." The latter, he added, was "generally connected with loss in terms of costs and time."[25] Again, Gerhard emphasized that in addition to practical-technical knowledge, the mining official also needed generalized empirical knowledge, provided by geognosy, in order to be able to react fast enough to the manifold changes occurring at different mines and smelting works.

A mining official's suggestion for technological improvements profited from translocal empirical knowledge of what was doable, which resulted from the comparison of various mines and smelting works, from knowledge transfer, and last but not least from chemists' experiments with metals and minerals. In contrast to geognosy, chemistry was a science with a long tradition that had already integrated metallurgical knowledge.[26] But only in the second half of the eighteenth century did "metallurgical chemistry" develop as a special subdiscipline. This internal differentiation process in chemistry occurred in the mining milieu, especially in the framework of the mining academies. Metallurgical chemistry dealt with the same materials and processes as the smelting techniques in smelting works. In chapter 8 we followed Gerhard's chemical-technological experiments that investigated the melting behavior of metals and the production of alloys. Such explorative experiments produced a body of empirical knowledge on metallic materials and their chemical behavior. While not directly transferable to industrial metal smelting, they delivered indications—a "guideline" (*Leitfaden*), as Gerhard put it—for what was doable. Gerhard thus pointed out that "an otherwise skilled but merely practical smelter, who comes to a new smelting works," would have to "work long and in vain before he discerns the process appropriate for this location, if he does not have a proper grasp of the chemical principles that could serve as a guideline for this purpose."[27]

Gerhard argued that mathematics, mechanics, and hydraulics were also indispensable components of mining science. Geometrical knowledge had long played a role in the surveying of mines and in the drawing of mine plans. Thus, the mine surveyors were regarded as practical mathematicians and land surveying as *geometria subterranea*.[28] For the new technology of mining machinery, additional mathematical and physical knowledge was needed, Gerhard reasoned, "since in mining, the surveys above and under ground need to be done, and no mine or smelting work is imaginable without machines that are driven by fire, air, or water"; further, "since both in mining and in smelting various calculations occur, mathematics, along with physics, mechanics, and hydraulics, is indispensable."[29]

Gerhard's plan thus made a series of concrete arguments for the usefulness of mathematical and natural-scientific knowledge for mining officials. He emphasized the usefulness of accumulated and generalized empirical knowledge, obtained through systematic inquiry in the context of scientific institutions. Classroom teaching of mining science linked the empirical knowledge of the man of science with the knowledge of the technical expert.

SCIENTIFIC AND PRACTICAL KNOWLEDGE IN THE CONTEXT OF MINING SCIENCE

In their attempts to create a discipline of mining science, its practitioners attempted to identify and compile all the existing building blocks of knowledge of potential relevance to mining and metal production. These included tried and tested technical knowledge of the construction of mines, mining machines, water drainage, mine surveying, ore extraction, assaying, and smelting techniques that together comprised mining technology, as well as parts of mathematics, mechanics, hydrostatics, mineralogy, chemistry, and economics. Beyond the bundling of such existing components, the science of mining also fostered the creation of new fields of knowledge, such as geognosy, physical geography, metallurgical chemistry, and mechanical engineering (*Maschinenlehre*). Thus, both Heinitz and Gerhard defined mining science as a complex of natural-scientific, technological, mathematical, and economic knowledge. Mining science was considered to be useful for mining officials, both the technical officials involved directly in mining and metal smelting, and high officials such as mining councilors and mining masters. In addition to teaching, mining science also comprised research, in particular in geognosy, land surveying, map making, studies of materials, as well as mechanical and hydrostatic studies of mining machinery.

In the historiographical literature on eighteenth-century mining, mining academies have been a prominent object of discussion, while studies of the discipline of mining science are relatively rare. It has often been ignored that early modern mining science was broader in scope than modern mining science and that it included components of mathematics and of natural sciences such as geognosy, physical geography, mineralogy, metallurgical chemistry, hydrostatics, and mechanics. One issue, however, has been frequently discussed: the relationship between scientific and practical knowledge within mining science. Referring to the *"science des mines"* in France, Isabelle Laboulais has emphasized that mining science should be understood as hybrid knowledge (*savoir hybride*) that was anchored "in natural history, the physical sciences, and technology."[30] In a lecture at the school of mines in Paris, established in 1794, this was formulated thus: "We will never lose sight of the fact that the engineer and the director of the mines have their place between science and art and that they have the task of establishing the most intimate connection between them."[31] Christoph Bartels comes to a similar conclusion with respect to mining science in the German states. Regarding the introduction of the water column machine, he has pointed out that it was "characterized by increasing mathematical and scientific-experimental interpenetration of (hydro)mechanical processes used for mining purposes."[32] In his analysis of Henning Calvör's textbook on mining machines from 1763, which he characterizes as a "scientific-technological" textbook, Bartels concludes that it was based "on the theoretical tools of its time *and* the experience from operational practice."[33] He also argues that the experts of mining engineering sought to connect scientific, technological, and practical-technical knowledge.

While these authors rightly point out that eighteenth-century mining science brought natural-scientific, mathematical, technological, and practical-technical knowledge into close interaction, other historians have driven a wedge between the scientific and practical components of mining science. One extreme is the older assertion that advances in eighteenth-century German mining technology were caused solely by the "application" of mathematics, physics, and chemistry.[34] Another extreme is the more recent view that the protagonists of mining science "rejected practical knowledge," and that there was a "fundamental lack of understanding of the mechanisms of practical-artisanal knowledge culture on the part of contemporary science."[35] According to the latter view, talk of mining science was merely rhetorical and the expression of an ambition for power on the part of scientifically educated mining officials. In a similar vein, it has been argued that the miners only gradually and reluctantly accepted resources from the sciences because they stuck to their "traditional and tested methods."[36] Apart from ignoring the fact that mining science addressed mining officials rather than ordinary miners, this argument posits "time-tested methods" as the principal alternative to new procedures introduced with the help of the mining sciences. However, reformers frequently introduced novel production techniques that coexisted alongside older ones, as will be detailed in the next chapter.

12 THE SCIENCE OF SALTWORKS

In the winter of 1791, while still a student at the Freiberg Mining Academy, Alexander von Humboldt wrote a long essay on salt making and the science of saltworks (*Salzwerkskunde*). Humboldt's essay yields further insights into the contemporary understanding of useful sciences. The text, which was published in the *Miner's Journal* (*Bergmännisches Journal*) in 1792, was part of Humboldt's preparation for his future duties as a Prussian mining official, which would include the inspection of saltworks. But it went far beyond this goal. In the last third of the eighteenth century, there were numerous publications on saltworks and ways toward their improvement.[1] Humboldt's publication differs from the majority of these texts in an important respect. Instead of concentrating on the improvement of traditional saltworks, its main focus is on how to introduce novel, knowledge-based production technologies, the knowledge here being of the chemical composition of common salt.[2] Moreover, Humboldt reflects on the aims of saltworks science and with that on the aims of useful science. With respect to the latter, it is unlikely that the text presents just the individual ideas of a twenty-two-year-old student of the Freiberg Mining Academy. More likely, Humboldt had absorbed ideas about useful science that were already in circulation in Freiberg and articulated these ideas further. As we will see, he described saltworks science as a particular branch of "technical chemistry." In parallel to the institutionalization of the useful sciences, internal differentiation in already existing disciplines was happening, such as the demarcation of "technical chemistry"—or of "metallurgical chemistry" and "pharmaceutical chemistry"—from the disciplinary core of chemistry.

Over dozens of pages of his *Salzwerkskunde*, Humboldt portrayed the time-tested artisanal operations in the German saltworks, which he had become familiar with on his travels and through additional reading.[3] He detailed the local conditions in the saltworks and the artisanal methods of salt extraction from brine—in Germany the dominant type of salt extraction, alongside the much less common mining of rock salt. Thus, he described the different graduation methods, the structure of graduation towers, their building materials, thickness, and spatial orientation, which changed according to wind direction and sun position. Further, he detailed the various local methods of salt boiling in pans, the second principal method of salt extraction from brine. He described the various types of pans, their size, form, and the materials they and their covering mesh were made of, and the furnace types, drafts, and smokestacks, as well as the kinds of fuel, such as wood, peat, or hard coal, and their use, which varied according to local conditions. Finally, he added detailed descriptions of the boiling process itself. All this reads like a hymn to the arts and crafts, culminating in the following remarkable statement: "The empirical salt-maker [*Halurge*] will continue on the path taken, with a sure step, without any scientific knowledge."[4] The long, descriptive part of Humboldt's essay demonstrates compellingly that the aim of the reformers in mining administrations, in Saxony and in Prussia, was not a wholesale displacement of practitioners' time-tested technical knowledge by scientific knowledge, but rather the targeted use of scientific knowledge for improving specific parts of existing technology and for introducing new ones.[5]

Another part of Humboldt's essay presented elaborate reflections on the meaning of useful science. Humboldt claimed rather ambitiously that his proposals would put to rest "the conflict between the theoretician, who often so impetuously offers his advice unsolicited, and the practitioner, who often deliberately avoids it."[6] In a brief discussion of the contemporary physical theories of vaporization and evaporation, he first made clear that this kind of high theory did not contribute to saltworks science or help to solve the conflict between the "theoretician" and the practitioner. Rather, the insights of physics and chemistry "into the causes of vaporization" and the chemical process of evaporation were too abstract to meet the demands of practice. With respect to practice, there were many additional things to consider, especially "hindrances based in the local conditions or other secondary conditions," that, Humboldt added, "can vary in many respects according to the nature of the locality" and therefore cannot "be expressed in general, useable formulas." Instead, he continued, they "must be examined in individual cases, in accordance with tried and tested economic and physical experience."[7] These were not the words of a man who believed in the omnipotence of natural science. They rather highlighted the need for a new type of science that coupled knowledge components taken from the natural sciences with generalized technological and local technical knowledge.

Humboldt believed that for the science of saltworks, the empirical parts of contemporary chemical science and chemical analytical methods were more useful resources than high physical theory. Right at the beginning of his essay, he defined the science of saltworks as a special branch of "technical chemistry," a term appropriated from the physician and chemist Johann Friedrich Gmelin (1748–1804), who taught at the University of Göttingen and whom he had met during his studies in Göttingen from spring of 1789 until spring of 1790.[8] In his textbook *Principles of Technical Chemistry* (*Grundsätze der technischen Chemie*) Gmelin portrayed technical chemistry as an endeavor that proved that "learned knowledge" had an "influence on the arts and industry of civil life." It did so, Gmelin explained, not by imposing learned knowledge on practice, but rather by bringing "the two closer together."[9] In the same vein, Humboldt set out to argue that his saltworks science created a bridge between learned chemical knowledge and the practical-technical knowledge embedded in saltworks.

Recent chemistry had shown, he wrote, that common salt consisted of an alkaline component, the "mineralogical *Laugensalz*" (alkaline salt), and "*Küchensalzsäure*" (muriatic acid, later termed hydrochloric acid), in a specific proportion. Older, divergent opinions on its composition were "sufficiently refuted," and the available chemical-analytical knowledge was clearly sufficient for technical purposes. In Humboldt's words: "With the current state of chemistry, our knowledge of the composition of common salt is precise enough for the demands of technical works on a large scale."[10] But what concrete consequences did the chemical knowledge of common salt have on the practice of salt making? Humboldt did not argue here for the rationalization of traditional salt production, but rather for something completely different: the invention and production of new useful substances. He proposed that one could chemically separate common salt on a large scale, in order to isolate its alkaline component and its "salt acid" component. Then one could further process the salt acid (muriatic acid) and extract its gaseous reaction product, "dephlogisticated muriatic acid" (later termed chlorine). By the time of the publication of his *Salzwerkskunde* (1792), the proposed processes had already been developed. In the 1790s in France and England, chlorine produced by chemical decomposition of common salt was used as a new kind of chemical bleach in the textile industry, and the industrial application of the Leblanc process for soda production was on its way.

INVENTION OF NEW PRODUCTION TECHNOLOGIES

The fact that common salt contained a "spirit of salt" (or salt acid, muriatic acid) was already well known in late medieval alchemy, but it was not until the seventeenth century that Johann Rudolph Glauber (1604–1670) published the various methods of its isolation from common salt. In 1774 the Swedish chemist Carl Wilhelm Scheele (1742–1786) described the isolation of chlorine gas from muriatic acid and manganese. He interpreted the chlorine gas as "dephlogisticated muriatic acid" and observed that it destroyed plant pigments—an observation upon which Humboldt had based experiments on plants in Freiberg.[11] Taking up Scheele's observation, in 1785 the French chemist Claude-Louis Berthollet (1748–1822) undertook the first large-scale technological experiments on chlorine bleach, the results of which were adopted by French and English companies. For use as bleach, chlorine was dissolved in a fluid, for instance in a liquid containing potash, which was marketed as *Eau de Javel*.

Moreover, on September 25, 1791, thus right at the same time that Humboldt was writing his *Salzwerkskunde*, the Leblanc process for producing soda (sodium carbonate) from common salt was patented. In 1736, the French chemist Henri-Louis Duhamel du Monceau (1700–1782) reported to the Paris Academy of Sciences about his analyses of common salt. He had been successful in isolating an alkaline component, the *"sel alcali fixe,"* and had proved that the same component was also contained in soda.[12] His subsequent discovery of two reaction procedures for the production of soda from common salt demonstrated that soda could be made in a new, artificial way. But a significant task had yet to be mastered: the transfer of the chemical experiments on the laboratory scale to a cost-effective industrial procedure. From the early 1770s on, several English and French chemists and entrepreneurs dedicated themselves to this problem, with the method developed by Nicolas Leblanc (1742–1806) finally winning the race. In the fall of 1790, even before the patenting of the Leblanc process, the first soda factory was built in Saint-Denis, some ten kilometers north of Paris.[13]

Humboldt was familiar with the commercial production of "dephlogisticated muriatic acid" (chlorine) and its use as a novel bleaching agent. "Mr. Berthollet," he observed, was "gracious enough … to make public his whole bleaching process, including Mr. Decroizille's suggestions for improvements." He proposed that this invention be put to use in Prussia, and that the Prussian "saltworks might begin to use an invention that would advance the flourishing of the linen and cotton manufactories and thereby as well the welfare of the nation's industrious and most needy classes."[14]

To this end, in the immediate vicinity of the saltworks—and in addition to the production of salt—he proposed that new chlorine production plants be set up. In contrast to chlorine production, the Leblanc method of soda production was still unknown to Humboldt. Thus he merely pointed out that "the decomposition of salt for the use of its alkaline component could be of great significance to many saltworks."[15] He recounted the experiments hitherto conducted on the decomposition of salt and the isolation of its alkaline components, which, however, he considered "not suitable and simple enough for preparation on a large scale."[16] But here too, the aim of his proposal is quite clear. Further technological inquiry was needed to improve the techniques of isolating the alkaline component of common salt on a large scale. Based on this, a new branch of production should be attached to the saltworks, thereby procuring them an economic advantage. "For Germany's trade balance," wrote Humboldt, "the domestic fabrication of mineral alkali would be highly desirable."[17]

Humboldt's main concern was to expand the salt-making industry through technological inquiry and new innovative production technologies. In conclusion, he generalized his statements on the usefulness of analytical chemistry by mentioning two further methods that had already been introduced: silver isolation via the Bornian amalgamation method, with which he had become acquainted in Freiberg; and the industrial production of sulfuric acid from sulfur using the lead chamber method developed by John Roebuck in the middle of the eighteenth century. "How important analytical chemistry can be for industry and the wealth of nations [*Nationalreichtum*]," he wrote, "is evidenced by the amalgamation process, the preparation of sulfuric acid from sulfur, and the above-mentioned bleaching by dephlogist[icated] muriatic acid."[18]

In Humboldt's view, the use of chemistry lay thus not in high theory, but rather in the production of conceptual and empirical knowledge of the composition of substances and methods of their chemical analysis.[19] According to him, chemistry was important for the technical expert (*Techniker*) because "it familiarizes him more closely with the components and properties of the intended product."[20]

MAKING THE CASE FOR TECHNOLOGICAL RESEARCH

Toward the end of his essay, Humboldt explained his vision of a science of saltworks:

> The scientific treatment of an art follows a method that is itself completely distinct from its practical performance. The former needs to collect principles from

all related sciences, compare the experiences of the physicists with technical experiences, and observe each apparently unimportant circumstance. However, the latter, [that is,] the method of practice, must take the opposite course if it is not to fail in its aim or lead to timid indecisiveness. Once a method has been selected, practical performance must isolate itself, as it were, dealing with just one object, and take local conditions more into consideration than general speculations, and not let itself be distracted by small circumstances, sacrifice minor advantages for larger ones, etc.[21]

This passage draws a clear distinction between the institutionalized useful sciences and the technical knowledge and operations embedded directly in industrial practice, or, in Humboldt's words, between the "scientific treatment of an art" and its "practical performance." The "practical performance" of an industrial art stood under the pressures of time and economic success. It had to deliver products and thus make decisions quickly on the appropriate means to this end. Instead of exploring new and unexpected phenomena that cropped up in the course of inquiry, it had to concentrate on the production of a commodity, "isolate itself, as it were, dealing with just one object," as Humboldt put it. Singleness of purpose and knowledge of "local conditions" were indispensable for any practical success.

For Humboldt, the "scientific treatment" of an art followed a very different method. The protagonists of the useful sciences needed to identify and select the potentially useful parts of the natural sciences and link these to technical experience. From the comparison of scientific and technical knowledge resulted a creative tension that led to new insights. Moreover, close observation of all phenomena, including perplexing, "apparently unimportant circumstances," was crucial for the success of the useful sciences, as it might open up new horizons of knowledge. Further, the useful sciences generalized their observations rather than focusing on "local conditions," and could even afford to make "general speculations." All this is nothing other than the description of research. An important factor for research, Humboldt emphasized, was absence of time pressure and freedom to explore things.

We can sum up Humboldt's understanding of *Salzwerkskunde* thus: *Salzwerkskunde* should be pursued as a science and, in contrast to industrial practice, allowed the time and space for time-intensive, open-ended inquiry. *Salzwerkskunde* was thus defined as a particular type of science which developed its own forms of research, including work on invention. While its immediate aim was production of knowledge, its ultimate aim was the improvement of industry. Thus, the idea that saltworks science was "useful" did not imply that it was attuned to immediate application.

As Humboldt's examples show, there can be intricate detours to usefulness. Like mining science, saltworks science belonged to a third sphere located between the natural and mathematical sciences and the world of practical-technical knowledge and industry.

Humboldt's essay demonstrates clearly that, already in the late eighteenth century, there was an awareness among reform-minded officials and men of science of the existence of an important difference between a "science" relating to technology and technical knowledge directly embedded in industry. This difference needs to be emphasized, especially in view of a trend in the English-language literature that subsumes under "technology" technical things and procedures, practical-technical knowledge, as well as the engineering and technological sciences established at distinct institutions. For example, in a recent handbook on "engineering science," the philosopher of science Hans Radder argues that it does not make much sense to differentiate between technology and technological science, because both pursue the same goal of technological improvement.[22] In line with this, the handbook deals predominantly with technology in the sense of technical things and technical knowledge embedded in industry.[23] This approach not only ignores the factual existence of distinct institutions for the teaching of and research in technology, which from the second half of the nineteenth century existed in the English-speaking world as well. It also implies the questionable epistemological assumption that human practices are fully determined by the actors' goals. Although practical-technical knowledge and the technological sciences shared the aim of industrial improvements, their respective ways and means of achieving this aim are by no means identical, and at times it is the ways and means that are key to success. Humboldt's differentiation between the "practical performance" of an art and its "scientific treatment" was not motivated by an appetite for polemic. Only with institutionally anchored useful sciences were the spaces created for sustained technological research and teaching and thus the sustained organization of industrial innovation. We will discuss these issues further in chapter 17.

HISTORICAL ROOTS OF USEFUL SCIENCE

Attempts to understand how technical devices work, and the natural conditions for their construction, were by no means a novelty of the second half of the eighteenth century, introduced with the German useful sciences and the French engineering sciences. Rather, such attempts can be traced back to the Renaissance and even the late Middle Ages. They were rooted in the challenges presented by new technical and commercial endeavors, such as the construction of fortifications, mining operations reaching to great depths, the expanding trade in colonial products, and the market for luxury goods, which exposed the limits of practical knowledge passed down for generations in the system of apprenticeship and created a demand for a new stratum of innovative and experimenting engineers and other technical experts. From the sixteenth century onward, technical experts began to publish their knowledge. In their technological books, they described their inventions just as much as they extolled the usefulness of their instruments and machines. They also collected practical-technical knowledge, scattered over many different locations, and created comprehensive technological descriptions of the instruments, techniques, and know-how that translocally pertained to specific practical fields (see chapter 5).

Writing such summaries of translocal technologies, or, in Bacon's terminology, "experimental histories," required abstraction from numerous tiny features that were specific to local practices, and identification of their shared characteristics.[24] The authors of early modern technological books also occasionally reflected on artifacts and techniques as well as their natural conditions. In so doing, they sometimes employed elements of mathematics, astrology, alchemy, mineralogy, and other scientific disciplines of their time. Georg Agricola's *De re metallica* is a famous example of an early technological book that went far beyond the boundaries of local technical knowledge and offered further ad hoc explanations that included learned natural knowledge.[25]

There were also mixed technological-scientific publications, such as the seventeenth-century chemical treatises and textbooks that contained "practical" as well as "theoretical parts." While the former collected recipes about chemical-pharmaceutical preparations from many different places, the bulk of their "theoretical" parts consisted of knowledge about the properties and relationships of material substances. The overriding question tackled by the authors of these books was what it was possible to discover, invent, and make in their physical environment, and what kind of knowledge lent power to mankind to manipulate transformations of material substances. The early modern chemists considered chemistry to be both "science and art." They believed that studies of natural things affected the chemical arts and, conversely, that studies of chemical preparations could illuminate chemical operations in nature. In many of their studies, the ancient distinction between natural things and artifacts became irrelevant.[26]

In the early seventeenth century, the English philosopher and statesmen Francis Bacon (1561–1626) gave voice to this approach to knowledge. In his *Novum Organum* (1620) he proclaimed that "human knowledge and human power meet in one." Bacon envisioned sciences (*scientiae*) that could "help us in finding out new works," because without insight into nature it was impossible to create new useful things. Knowledge about nature would lead to "methods of invention" and new "works," because "that which in contemplation is as the cause is in operation as the rule."[27] In line with this, Bacon argued that studies of inventions and technical operations "take off the mask and veil from natural objects." He thus developed a program of an "experimental history" that comprised both the "history of arts, and of nature as changed and altered by man." The core of experimental history, he pointed out, was formed by those arts that "exhibit, alter, and prepare natural bodies and materials of things; such as agriculture, cookery, chemistry, dyeing," and so on, "because many things occur in them which relate to the alterations of natural bodies."[28] Thus, Bacon argued for the systematic coupling of natural and technological inquiry. His "science" was not merely a science of nature but also of human arts and crafts, or technology. In his *New Atlantis* (1627), he described a community of researchers on a fictional faraway South Sea island, "Atlantis," who combined natural and technological inquiry and put their science at the service of the common good.

The Royal Society and other scientific societies adopted Bacon's concepts of practical experience and useful knowledge. Eighteenth-century cameralism and parts of the Enlightenment movement propagated these ideas further. A wave of popularization of natural-scientific and technological writings began, borne by the idea that the education of broad swaths of the population would advance technological and social progress. Through public lectures and the publication of easy-to-understand texts, associations such as the Lunar Society in Birmingham supported the spread of natural-scientific knowledge alongside knowledge about the arts and crafts. Scientific societies on the European continent, such as the economic and patriotic societies, pursued similarly broad aims of enlightenment and education.[29]

With the incipient industrialization and rise of modern state bureaucracies, the discourse on useful knowledge underwent a historical transformation, and with it the role of experts. Terms like "*science de l'ingénieur*" and "useful science" epitomize this transformation. New disciplines of mixed technological-scientific knowledge emerged along with new knowledge institutions such as the French engineering schools and the mining academies of Freiberg and Schemnitz. Mercantilist policy and demands of the state administration were the immediate factors promoting these changes, as the useful sciences were aimed not at craftsmen per se, but rather at engineers and other technical experts who were, as a rule, state officials.

CAMERAL SCIENCE AND "TECHNOLOGY"

Mining science, saltworks science, and other useful sciences established in the last third of the eighteenth century were intended for the education and training of officials of the mining administrations and the new technical departments in the state bureaucracy established in Prussia after the Seven Years' War. Earlier in the century, the German universities had introduced the so-called "cameral science" and "technology," which had similar functions to the useful sciences but also differed from them in important ways.[30] The university discipline of cameral science involved a concept of useful knowledge that was also oriented at the requirements of state officials. However, cameral science was significantly broader in scope than the useful sciences, as it compiled knowledge drawn from social, political, and economic theory, fiscal and legal knowledge, rules of administration, as well as some mathematics and natural science. At the Prussian University of Halle, cameral science was introduced in 1727, that is, in a time prior to the establishment of the specialized technical departments of the state bureaucracy. It addressed only leading officials concerned with a broad range of fiscal and administrative matters. Cameral science prepared leading officials mainly for paperwork, with natural and technological knowledge functioning as auxiliary means for fiscal and administrative tasks.

In 1777, Johann Beckmann (1739–1811), professor at the University of Göttingen, published a textbook on technological knowledge, titled *Introduction to Technology (Anleitung zur Technologie)*. "Technology" in Beckmann's sense was systematically compiled descriptive knowledge about materials, techniques, and artifacts embedded in the contemporary arts and crafts, including agriculture. Materials, both raw and processed, were highlighted as a key factor interconnecting agriculture and the various arts and crafts. Even though Beckmann's "technology" was narrower in scope than cameral science, it covered the arts and crafts broadly. Thus, it was not a discipline intended for the professional training of particular groups of technical officials, as was the case with the useful sciences. Furthermore, Beckmann's technology was still in the mold of the descriptive tradition of technological books and Baconian "experimental history," regardless of the fact that it also had some explanatory functions.[31] In his 1806 *Outline of General Technology (Entwurf der allgemeinen Technologie)* Beckmann proposed to develop a systematics of the functions of labor, and to supplement it with a systematics of the appropriate instruments and machines. This was a step further toward analysis and understanding, but it was still primarily a way a way of ordering the teaching material.[32]

Cameral science and technology differed in several respects from the useful sciences. First, the former were just teaching disciplines that did not stimulate in-depth analyses and additional inquiry comparable to the experimental research carried out in the framework of the useful sciences. Second, in contrast to cameral science and technology, the useful sciences provided an education and training tailored to the professional demands of expert officials working in the technical departments of the state administration. Third, the useful sciences were anchored in new types of institutions, such as mining academies, that facilitated the interconnection of theory and practice.

13 THE FIGURE OF SCIENTIFIC-TECHNOLOGICAL EXPERT

The useful sciences were promoted by Prussian ministers and men like Werner, Klaproth, Achard, Gerhard, and the young Alexander von Humboldt. These men were inventors and technical experts, but they were also recognized as scientists (*Naturforscher*). They were members of scientific societies and academies, experimenters, authors of scientific texts, and organizers of schools for the useful sciences. We designate such persons, who stood with one leg in the republic of letters and one in the world of technology, as scientific-technological experts.[1] Many of them were state officials, who also possessed fiscal and administrative knowledge.[2]

The scientific-technological experts defy our attempts at unambiguous categorizations because they unify in their person roles and forms of knowledge that conventionally were distributed across disparate social groups and professions. The study of their social origin and education also does not allow for their classification within the traditional knowledge economy. For example, Klaproth came from a humble artisanal background, had completed an apprenticeship, was an apothecary by profession and a self-educated chemist, who advised Minister von Heinitz on chemical-technological issues. Gerhard came from a middle-class family, had studied medicine, and was trained as an expert official as well as a mineralogist and chemist in the milieu of the mining administration. Humboldt, whose family belonged to the lower nobility, had studied briefly at university, attended the Freiberg Mining Academy, and then improved both his natural-scientific and technological expertise during his time as a mining master.

Experts, regardless of what color, were bridge builders between the worlds of labor and commerce on the one hand and the world of science on the other, and the bridges were built from both sides. If the figure of the scientific-technological expert is the focus of attention here, this is due to the fact that this person has specific characteristics which caused his recognition as both *Naturforscher* (man of science) and technical expert.

Using their professional environments as laboratories for observation and experimentation, the scientific-technological experts brought technological and scientific inquiry into a mutually beneficial relationship. Further, they were the main practitioners of the useful sciences. It was primarily the scientific-technological experts who participated in the discourse on the useful sciences in Prussia. Along with ministers and other high-ranking state officials, they also worked to create the institutions of useful science. They elaborated the first programs and curricula for this new type of science, and they were also teachers at mining academies and similar technological schools, in addition to the engineers and technical experts. In the framework of these schools, the scientific-technological experts were the main researchers or organizers of research as well. Because the development of the useful sciences and the specialized natural sciences went hand in hand, the scientific-technological experts were also among the founders of new scientific disciplines and subdisciplines such as geology, chemical mineralogy, and analytical chemistry.

Like the early modern engineer, inventor, and technical expert, discussed in chapter 5, the figure of scientific-technological expert has a deeper history. Recent historical research has drawn attention to the numerous areas of partial overlap between natural inquiry of learned men and the practices and knowledge of inventive artisans in the Renaissance and early modern world.[3] The Renaissance engineer, who associated with learned men, wrote mathematical and technological texts, and created impressive artworks, has long been a familiar figure. As Pamela Long has pointed out, zones of sustained collaboration between Renaissance engineers and academically trained men brought forth a third figure: one who united the roles of the learned man and the engineer or inventive artisan. Hence, Long states, "the wide divisions between workshop-trained artisans and university-educated scholar/humanists narrowed *and in some cases disappeared.*"[4] As examples, she mentions Leon Battista Alberti (1404–1472) and Lorenzo Ghiberti (1378–1455).

To mention a few more examples, in a recent study on Simon Stevin (1548–1620), who was renowned for his work on hydrostatics, Alan Chalmers has pointed out that Stevin was both a mathematically trained engineer and a scholar, or in other words a "scientist/engineer."[5] A similar case is that of Galileo Galilei (1564–1642), celebrated by most historians solely as a great scholar. As has been shown in a project carried out at the Max Planck Institute for the History of Science, Galileo was also a practical mathematician and owner of a workshop, where he constructed mathematical instruments and taught practical mathematics. He also actively contributed to the solving of engineering problems at the Venetian Arsenal. As his activities in practical mathematics/engineering and theoretical mechanics significantly intersected, Peter Damerow and Jürgen Renn have called him a "scientist-engineer."[6]

A similar example is that of the "English Galileo" Thomas Harriot (1560–1621). As Matthias Schemmel has shown, Harriot was a practical mathematician who worked on projectile motion and the free fall of bodies in no less sophisticated ways than his contemporary Galileo. Comparing him with Galileo, Schemmel comes to the conclusion that "practical mathematicians attempted to solve practical problems in navigation, surveying, shipbuilding, fortification, gunnery, and similar fields of practical concern.... They were, however, distinct from the majority of practitioners by their reflection on the practical knowledge. They strived for a mathematization of that knowledge, they edited ancient works on mathematics and translated them into the vernacular," and they also performed experiments and authored manuals and books.[7]

Among the chemists of the seventeenth century, there were numerous apothecaries who had completed a pharmaceutical apprenticeship and came into contact with the academic world. Some of them participated in learned chemical discourse, not only sporadically but on a more permanent basis. They began to experiment in a more systematic and sustained way, read books, authored chemical texts themselves, and were eventually recognized as chemists as well.[8] For example, with respect to the apothecary and alchemist Rudolf Glauber, Pamela Smith came to the conclusion that he was both a man of science and an artisan.[9]

In the eighteenth-century mercantilist states such as Saxony and Prussia, the state administration hired scientific-technological experts for long-term public service. In eighteenth-century England, the model country of industrial capitalism, we encounter comparable figures primarily in the artillery and navy, while in the civilian sector the scientifically knowledgeable private entrepreneur stands out, which the English historian Peter Jones has called the "savant-fabricant."[10] The eighteenth-century British savant-fabricants and the scientific-technological experts in continental Europe had many characteristics in common, particularly their interconnection of inventive work and natural-scientific inquiry. They also shared engagement for the nation-state and for the common good. Almost all of these men participated in patriotic movements seeking to foster technical progress and the industrialization of their country.

In the decades shortly before and after 1800, the Prussian scientific-technological experts created strong links with the state bureaucracy. This changed in the course of the nineteenth century in parallel with the

economic withdrawal of the state and the transition to a liberal industrial capitalism. In Prussia, the figure of savant-fabricant appeared first in
the nineteenth century; an example is Werner von Siemens (1816–1892),
who was at once a fabricant, inventor, physicist, and member of the
Royal Prussian Academy of Sciences.[11] At the end of the nineteenth century, scientific-technological experts were often employees of the German
chemical and electrotechnical industry as well as teachers at the technical
universities, or researchers at the newly founded technological-scientific
research institutions such as the Imperial Institute of Physics and Technology (Physikalisch-Technische Reichsanstalt) in Berlin.[12] Generally speaking,
in the second half of the nineteenth century there were still engineers and
scientists in Europe who frequently crossed the borders between those two
communities. The historian of technology Edwin T. Layton has designated
these intermediaries as "engineer-scientists."[13] Engineer-scientists were the
most important successors to the scientific-technological experts who had
been working in the milieu of state-directed manufacture and bureaucracy.

DISCUSSION I: THE "SCHOLAR-AND-CRAFTSMAN" THESIS

Achard, Klaproth, Werner, Gerhard, and the young Alexander von Humboldt have been presented as *scientific-technological experts* who united the
roles of both men of science and inventive technical experts. The term
"scientific-technological expert" should be understood not least as a clarification with respect to the older "scholar-and-craftsman" thesis. The height
of the Cold War in the 1950s saw a passionate debate among historians of
science on the role of artisans and scholars in the Scientific Revolution.
The debate was ignited by the rereading of an essay by the historian and
sociologist Robert Merton and of the writings of Marxist historians and
philosophers such as Edgar Zilsel, Boris Hessen, and Henryk Grossmann.[14]

Merton had argued that the study of new technical objects and problems was an important stimulus for the Scientific Revolution in the late
sixteenth and seventeenth centuries, and that the knowledge of men of science and inventors had so often been in broad agreement that the two were
sometimes indistinguishable. Stated more pointedly, "the inventor and scientist were often one."[15] Rupert Hall, a defender of a more internalist view
of the Scientific Revolution, criticized this thesis and gave it a polemical
turn. Ignoring Merton's efforts at differentiation between different kinds
of practitioners, he wrote: "This seems to me to be the defect of the view
that sees the new scientist of the seventeenth century as a sort of hybrid
between the older natural philosopher and the craftsman."[16]

Thus the "scholar-and-craftsman" thesis was born, which Merton did not formulate as such, for good reason. Rather than asserting that the traditional scholar and the traditional craftsman—who arguably are two disparate figures—united their social roles, he had a very specific group of practitioners in mind: engineers, gunners, land surveyors, instrument makers, and all kinds of technical experts or inventors. Technical experts performed handwork, but the special nature of their knowledge, their investigative enterprises, and their communication with men of science distinguished them from ordinary craftsmen. They were rather inventive specialists with extraordinary abilities for tackling difficult technical tasks beyond the scope of the usual craft professions. The knowledge repertoire of the guilds was only one source from which they drew their expertise. Other sources, as we have seen, were books, conversations with men of science, as well as travels. Continually on the lookout for new knowledge, technical projects, and inventions, these men were also excellent experimentalists. Likewise, Merton did not focus on the traditional "scholar" but rather on the new man of science, who conversed with inventors and emphasized the role played by experience in the acquisition of knowledge. And he rightly observed the occasional hybridization of the roles of inventor and man of science. Therefore, it would be more appropirate to term this the "scientist-and-inventor" thesis, a formulation that would be substantively in agreement with our term "scientific-technological expert" as well as the term "engineer-scientist."

As Hall reiterated his criticism even more sharply in 1963, it is perhaps not too farfetched to assume that his reformulation of Merton's thesis pursued not only scholarly but also political aims: the stigmatization of Merton's thesis as vulgar Marxist and thus as an unscientific delusion.[17] The conceptual strictures thus imposed on the historiography of science still have an impact today, years after the end of the Cold War. A history of the historiography of science in the Cold War has yet to be written, but one thing seems to be clear: the counterpart to the anti-Marxist thrust of Hall's argument was the construction of the figure of the scientist as performing exclusively pure science. The modern scientist in the Western mold had to embody rationality and objectivity—and only these qualities. In John Krige's words, pure science was a "sure antidote to communism."[18]

DISCUSSION II: GILLISPIE'S QUESTION

In early modern France, the military was the first branch of the state to recruit scientific-technological experts, specifically those who had been educated and trained in engineering schools. The military engineer Bernard

Forest de Bélidor (1698–1761), for example, author of several books, was inducted into the Paris Academy of Sciences in recognition of his scientific accomplishments. Charles C. Gillispie thus concluded that his "life mingled the scientific and the military."[19] In the second half of the eighteenth century, men who were both technical experts and savants were also active in numerous areas of state-managed French industry. In *Science and Polity in France*, his classic 1980 study on the sciences and the state in the *ancien régime*, Gillispie thus poses an important question: Did the French savants deliberately make connections between natural inquiry and technical work, and was there a substantial interaction between their scientific and technological inquiry? Or, alternatively, did the French state more or less compel them to carry out technological projects and become advisors to the state?[20]

In an older study from 1957, Gillispie had clearly demarcated the scientist from the practical man. Quoting Lavoisier, he observed, "the spirit which moves the scientist is fundamentally unlike that which animates the artisan"; and regarding the members of the Paris Academy of Sciences in general, he argued that "technological *expertise* was at best their corporate duty." Correspondingly, he distinguished between "true scientists" such as Claude-Louis Berthollet and Gaspard Monge, on the one hand, and chemists such as Jean-Antoine Carny, Nicolas Deyeux, Bertrand Pelletier, and Jean-Antoine Chaptal on the other, "whose instincts and interests lay in production and enterprise."[21] By contrast, in *Science and Polity in France*, he pointed out that the Paris Academy of Sciences inducted not only scholars but also apothecaries and military engineers into its ranks.[22] In the concluding discussion he then approached the question, posed above, concerning the motivation of the men of science. A colleague once asked him how the savants themselves judged their state-commissioned technical projects and "how they felt about these duties": "How did they see themselves in relation to knowledge, industry, and government?" In his answer, Gillispie referred to his previous elaborations on Berthollet, whom he characterized as a scientist as well as as a man who was committed to the state and to the Gobelins (a manufactory of tapestries in Paris); and he then generalized: "I think that what was said of Berthollet in this respect is largely true of all, namely that he would not have divided his life into these three compartments."[23] The answer to Gillispie's question thus seems to be clear: the French savants themselves did not separate their activities into scientific activities on the one hand and industrial and state activities on the other.

Let us briefly relate Gillispie's question to the scientific-technological experts in Prussia. As Achard risked personal financial ruin through his technical projects, he is certainly not a good candidate for a man of science not truly motivated to undertake such enterprises. In the case of Klaproth, the constellation is somewhat different, as he profited to some extent from his activities in Heinitz's circle of experts, mainly through sales of chemicals to the royal manufactories. However, his engagement as a chemical expert for Minister von Heinitz—as a member of the inspection committee in the Royal Prussian Porcelain Manufactory, for example, or as the experimental reviewer of Achard's sugar beet project—went far beyond personal material benefits. Clearly, money was not his prime concern. On the contrary, Klaproth often renounced financial rewards, and this not only in the case of his invention of uranium yellow.[24] Social benefit may have been another personal interest, but when in 1786–1787 Klaproth joined Minister von Heinitz's circle, he was already an official in the highest Prussian medical authority, and thus had a path to advance his career. Furthermore, he possessed his own laboratory, was a teacher, and was on his way to becoming a respected man of science. Hence, in many respects the case of Klaproth resembles that of Berthollet. As the personal physician to Madame de Montesson and protégé of the duc d'Orléans, Berthollet had sufficient financial means and a private laboratory at his disposal, yet he took over the directorship of the Gobelins and began to do research on the production of pigments and textile bleach. According to Gillispie, Berthollet voluntarily went to work for the Gobelins with a clear goal in mind: to do "applied chemistry."[25] For Klaproth as well, the milieu of the Mining and Smelting Works Department offered new possibilities to interconnect chemistry, mineralogy, and technological investigation. The production of useful knowledge and the ideal of working for industrial progress and the common good were the driving forces in both cases.

DISCUSSION III: SHAPIN'S DEMARCATION OF "GENTLEMAN SCIENTISTS" FROM "INVISIBLE TECHNICIANS"

In his *A Social History of Truth* (1994), one of the most influential works of history of science of the 1990s, Steven Shapin distinguished clearly between the seventeenth-century figure of "gentleman scientist," such as Robert Boyle (1627–1691), and that of the "invisible technician," such as Robert Hooke (1635–1703), who collaborated with Boyle. Pointing out the unequal social status of the two men, Shapin observed that it was mainly Hooke who carried out the experiments in their collaboration. This had its terminological reflection, he stated, in the distinction between "knowledge" (referring to Boyle) and "skill" (referring to Hooke), which perpetuated, he argued further, the ancient distinction between scholar and artisan, theory and practice, the head and the hand.[26] As sympathetic as Shapin's concern about social inequalities is, his strong emphasis of this issue distracts attention from the quiet transformation going on in the knowledge economy of the time.

By contrast, in Merton's eyes Robert Hooke was a typical example of a hybrid scientist and inventor. He was "at once a considerable scientist and probably the most prolific inventor of his age," he pointed out.[27] In the seventeenth century, the ancient distinction between hand and mind had long been undermined at certain sites of practice. In the framework of state-organized technological projects, for example, engineers, inventors, and technical experts were fully recognized as producers of knowledge, and they did combine knowledge and skill (see chapter 5). Between the scholar and the artisan, a whole spectrum of new producers of knowledge was evolving, including the hybrid figure of scientific-technological expert. As Merton rightly emphasized, Hooke was not just an assisting technician but also an author, man of science, and member of the Royal Society.

Shapin's thesis has been widely accepted in the community of historians of science, and despite its Anglocentric perspective it has been transferred to other countries as well. Peter Dear, for example, has argued that the French savants of the Enlightenment perpetuated the old ideological separation of scholars and artisans by linking "knowledge" to written representation and thereby denying the artisans, who transmitted their knowledge orally, any claim to produce or possess knowledge. In the figure of D'Alembert he sees the typical representative of the French savant. Dear argues that in his preface to the *Encyclopédie* D'Alembert formulated the leitmotif for the ideological separation between head and hand; there "D'Alembert effectively makes a sharp demarcation between artisans and learned philosophers (and their readership), between head and hand."[28] The impact of ideology seems so effective that any further analysis of the actors' practice is moot. "This active creation of difference between domains claimed to be categorically distinct from one another," Dear thus observes, "is why the now-popular term 'technoscience' has such limited usefulness in historical work."[29] The term "technoscience" is an analytical tool that helps to focus historical analysis on the actors' practices rather than on ideology. In Dear's eyes, however, it necessarily leads to false history. This is a remarkable instance of ideological determinism, one that is questioned as much by Gillispie's studies as by more recent studies on the field.[30]

The thesis that the French savant of the eighteenth century internalized the ideological separation of mental and manual labor is already formulated in Roger Hahn's classic work on the Paris Academy of Sciences, *The Anatomy of a Scientific Institution* (1971). Hahn, however, offers an interesting alternative to Dear's interpretation of D'Alembert. He understands D'Alembert's preface to the *Encyclopédie* as the expression of the highest esteem for handwork. D'Alembert, Hahn states, argued vehemently against scholars' stubborn clinging to obsolete epistemic values, while at the same time explicitly excepting the artisans from such criticism and, on the contrary, portraying them as a model for emulation.[31] According to Hahn, D'Alembert went so far as to plead for a new fraternal relationship between the new man of science and the artisan. He adds, however, that D'Alembert was an exception in this regard. The "typical intellectual," he observes, "for all his good will and independence of mind, was unable to turn his back upon the weight of tradition that relegated the manual artisan to a lower echelon of society."[32] To what extent the Paris Academy of Sciences' exceptional position of power led some savants to construct an elitist self-image is an open question. The assumption is plausible, but it does not justify the argument that cooperation between savants and technical experts was undermined from the outset and made impossible in practice.

IV TOWARD NINETEENTH-CENTURY TECHNOLOGICAL SCIENCE

14 USEFUL KNOWLEDGE AT THE UNIVERSITY OF BERLIN

In 1810, in the wake of a patriotic resurgence after Prussia's defeat by Napoleon's army at the battle of Jena and Auerstedt, a university was founded in Berlin. Wilhelm von Humboldt (1767–1835), from February 1809 Privy State Councilor in the Ministry of the Interior and head of its Section for Religious Affairs and Public Education, forcefully argued for more efficient uses of Berlin's scientific establishments and for the founding of a university that would unite these establishments at the same location. This change of the institutional landscape had consequences for the useful sciences. The lecture series on mining science was canceled, and its natural-scientific parts integrated into the university curriculum. The trainees (*Berg-Eleven*) of the mining administration received tuition-free access to the university lectures on chemistry, technology, mineralogy, and physics. The university thus continued the teaching of future mining officials, but former attempts to create a discipline of mining science, oriented at the professional demands of mining officials, were abandoned. As we have seen in chapter 9, the useful sciences had been promoted by reform-minded ministers and officials in the General Directory, while Friedrich II had little interest in them. Well into the nineteenth century, the Prussian kings pursued a policy that avoided the foundation of new educational institutions and instead sought to economize by reorganizing existing ones. At first glance, the University of Berlin seems to be an exception to this rule. However, at the same time as the new university was founded, older institutions—the lecture series of the mining administration and the Medical-Surgical College—were disbanded. Furthermore, the Royal Prussian Academy of Sciences was instructed to share its chemical laboratory, physical cabinet, and observatory with the new university. Natural history collections, including the mining administration's mineralogical cabinet, were also moved to the university, which was another way to reduce expenses.[1]

It is commonly assumed that the young Berlin university was a modern university that promoted the pure sciences and humanities and enabled free teaching and research under one roof, in accordance with Wilhelm von Humboldt's neohumanist ideal of *Bildung*.[2] There are good reasons, however, to question this view. Over a hundred years ago, the university historian Max Lenz highlighted the dual character of the University of Berlin: on the one hand, Humboldt's neohumanist educational policy as well reform debates about university education influenced its faculty organization and teaching; on the other hand, it continued to educate civil servants.[3] The education of competent civil servants was even regarded as a paramount patriotic duty. Thus, the statutes of the university declared that its special mission was to educate "students competent for entry into the various branches of higher service in the state and church."[4] Whereas expertise in jurisprudence was a traditional requirement for high officials, the State Ministry's technical departments needed above all men with technological and scientific competences.[5] This was taken into account in the appointment of professors and through the inclusion of disciplines such as mineralogy, chemistry, physics, cameral science, technology, agricultural science, and forest science. Moreover, as we will see below, the mining administration exerted direct influence on the appointment of professors and closely cooperated with the Section for Religious Affairs and Public Education, responsible for the universities. Wilhelm von Humboldt's pointed formulation that the University of Berlin should pursue research in solitude and freedom was directed neither against university education of civil servants nor against the teaching of useful knowledge. Rather, it was a rejection of excessive ministerial intervention in universities' teaching activities, which had been tolerated for decades at the four Prussian universities in Halle (Saale), Königsberg, Frankfurt (Oder), and Duisburg. The next section provides a brief overview of these interventions.

MINISTERIAL INTERVENTION IN UNIVERSITIES—A SHORT HISTORICAL REVIEW

The Lutheran University of Halle had been founded in 1691–1692 not least for the advancement of "the common good."[6] Hence, it fostered mathematics, natural sciences, and, from 1727, cameral science. In 1756, the professor of mathematics Johann Joachim Lange, the successor to Christian Wolff's chair, organized a *Collegium Mineralogicum-Metallurgicum* for future mining officials, which dealt with the study of mine construction, mining machines, the assaying of ores, and metallurgical-analytical chemistry and smelting technology.[7] After the Seven Years' War, the Prussian state intensified its efforts to anchor useful knowledge at universities. In December 1768, the minister responsible for the universities, von Fürst und Kupferberg, directed the University of Halle to organize lectures on mining science.[8] About a year later, similar directives were issued for the instruction of mining science, mineralogy, and chemistry at all Prussian universities. In the summer of 1770, in parallel to Minster vom Hagen's initiative to found a mining academy in Berlin (see chapter 9), the minister reissued his directive, this time extending it to all four faculties of the universities with the order that students be urged "to apply themselves better to the sciences necessary for financial and cameral affairs."[9] Even theologians were tasked with the teaching of physics, natural history, and mathematics.

The universities bowed to this pressure. In March 1770, the University of Frankfurt (Oder) organized lectures on the "mining sciences" (*Bergwercks-Wissenschafften*), and announced that Professor Petrus Immanuel Hartmann would henceforth teach metallurgical chemistry (after Gellert), while Professor Georg Friedrich Curts would offer lectures on chemistry and mineralogy (after Lehmann).[10] In May of the same year, the University of Halle sent out a similar announcement about "lectures on mining, economic, and cameral sciences."[11] In a similar vein, the University of Königsberg offered lectures on cameral science at the law faculty, on mineralogy, metallurgy, and chemistry at the medical faculty, as well as on pure and applied mathematics, mechanics, hydraulics, civic architecture, natural history, and experimental physics at the philosophical faculty.[12]

The Berlin Mining and Smelting Works Department lent active support to these lectures. For example, after initial doubts whether the professor of philosophy in Frankfurt, Georg Friedrich Curts, was actually capable of explaining "the minerals present in the royal provinces," the department pledged to help and ordered Gerhard to send the "necessary reports" to the professor.[13] As Gerhard seems not to have immediately carried out this order, in February Curts complained that he had yet to receive "historical information" for his lectures on Prussian minerals, "foreseen for the kind pleasure of your Majesty." He added that his "most eager wish" was to "teach young people, who in the future can be usefully employed in Your Excellency's … Mining and Smelting Works Department."[14] Hence, Gerhard was ordered once more to send Curts the desired materials. Finally, the University of Frankfurt asked the Mining and Smelting Works Department to support the lectures through the donation of minerals: "It would be of great advantage to the academic lectures on natural history, mineralogy, and other sciences of service to mining," the letter went, if the Mining

FIGURE 14.1
The Prinz Heinrich Palace (mid-eighteenth century).
Source: Gandert 2004, 51.

and Smelting Works Department would "deliver all varieties of domestic minerals."[15]

Friedrich II approved these measures and supported them through an announcement in March 1770, which stipulated that candidates for positions in the mining administration who had been educated in mining science would be given priority. Candidates who "dedicate themselves to these sciences," the announcement stated, "should be considered preferentially in future provisions."[16]

USEFUL KNOWLEDGE FOR MINING OFFICIALS

On October 15, 1810, in the Prinz Heinrich Palace on Boulevard Unter den Linden, the University of Berlin—from 1828 the Friedrich-Wilhelms-Universität—commenced teaching. At the time, the Mining and Smelting Works Department was integrated into the Ministry of the Interior, as its Section for Mining and Smelting Affairs (Sektion für das Berg- und Hüttenwesen). This reorganization facilitated negotiations on the teaching of future mining officials, since the Ministry of the Interior also had authority over the Section for Religious Affairs and Public Education, which was responsible for the Prussian universities. Consequently, even while the dissolution of the lecture series on mining science was still being debated, two of the mining administration's lecturers were appointed professors at the university: Martin Heinrich Klaproth and Ernst Gottfried Fischer (1754–1831). Furthermore, the mining administration also intervened in the appointment of the professor of mineralogy, Christian Samuel Weiss (1780–1856).

Klaproth was one of the university's first professors, appointed professor of chemistry in early 1810. Wilhelm von Humboldt had proposed him to the king with the following words: "He has enriched his science with true discoveries, through which he earned an international reputation comparable to that of only a few other scholars in your Majesty's states."[17] After the university had opened its doors in the fall of the same year, Klaproth used the auditorium and laboratory of the Royal Prussian Academy of Sciences for his university teaching. In continuation of the chemical lectures under the authority of the mining administration, which he had offered since 1784, he now also invited the mining administration's *Eleven* to his university lectures.[18] The Academy's auditorium and the chemical laboratory were located in the new laboratory building that had been erected in 1802 under Klaproth's supervision.[19] As Klaproth was the director of the academic laboratory, its dual use for both academic research and university teaching did not pose a problem.

After debates about the dissolution of the lecture series on mining science had been settled, Ernst Gottfried Fischer was appointed *professor extraordinarius* for experimental physics in 1811. In his case, the exchange between the Sections for Public Education and for Mining Affairs is particularly well documented in the archive. Fischer had studied theology as well as mathematics and natural sciences (*Naturlehre*) at the University of Halle. In 1776, he became a teacher, first at the *Pädagogium* in Halle, and in 1781 at the prestigious Gymasium am Grauen Kloster in Berlin, where he taught mathematics, physics, and Latin. In the 1780s he also tutored Wilhelm and Alexander von Humboldt in mathematics, *Naturlehre*, and ancient languages.[20] Having published several textbooks on mathematics and physics, and translated scientific books from French into German, he was inducted into the Royal Prussian Academy of Sciences in 1803. The following year, the mining administration hired him to teach physics to future mining officials, after Hermbstaedt had withdrawn from teaching on account of his new responsibilities in the Department of Trade and Industry.[21] Fischer was also one of the civilian directors of the War School (Allgemeine Kriegsschule), founded in 1810 by General von Scharnhorst.

In December 1810, Fischer wrote to the head of the mining administration that he was ready to continue teaching physics to the administration's *Eleven* either in connection with the new university or, alternatively, in the still existing framework of the "mining institute" and thus "completely independent of the university." However, he pointed out, the best option would be "if both the Sections of Mining Affairs and of Public Education could agree to bring the mathematical and physical instruction of the *Eleven* into connection with the university." In favor of the latter alternative, he immediately added: "as this already has been done regarding the teaching of chemistry and mineralogy."[22] Fischer also presented an economic argument for the latter proposal. This would be, he stated, "financially advantageous for both sides," but in particular for the mining administration, because the university budget had to cover most of the costs of the new professorship for physics. Fischer intended to offer a four-hour course each semester on "experimental physics" and a four-hour course on physical geography (*physikalische Erdkunde*) for all students; and further, to respond to the special interests of the mining administration, he wished to hold an examination course exclusively for the mining administration's *Eleven* in the winter. As this meant that he "would have more work to do for the university than for the Mining Institute," he believed that "it would not be wrong if the former [the university] carried four-fifths and the Mining Institute one-fifth of the costs."[23] His appointment as the university's *professor extraordinarius* for experimental physics shows that his proposal was approved.

In summer of 1810, the university hired Christian Samuel Weiss as the professor for mineralogy. Wilhelm von Humboldt had originally intended to appoint another lecturer of the mining administration's lecture series, the mining councilor and mineralogist Dietrich L. G. Karsten, to this position. However, Karsten died unexpectedly in May 1810. A former student of Abraham Gottlob Werner and well known as a mineralogist, Weiss had previously taught physics at the University of Leipzig. In October 1810, the Minister of the Interior, Friedrich Ferdinand Alexander zu Dohna-Schlobitten (1771–1831), wrote to the head of the mining administration, Johann Carl Ludwig Gerhard (1768–1835), the son of Carl Abraham Gerhard, to declare that the "commencement of lectures at our university has made it urgently necessary to arrange the handing over of the mineralogical cabinet to its new supervisor, the professor of mineralogy." However, Gerhard had different views concerning the fate of the mineralogical cabinet (or collection) and the role of professor Weiss. As the mining administration had long been responsible for the mineralogical cabinet, in his response Gerhard claimed that it "was the co-owner of the cabinet." Further, he argued that the mining administration "had not only the right to free access to [the use of] the mineralogical cabinet but also to recruit Professor Weiss as often as it believes to be in need of his advice."[24] Thus, Gerhard wanted to continue the tradition of employing scientists as state consultants.

Regarding the teaching of mineralogy, Gerhard insisted on a formal arrangement that would exempt the mining *Eleven* from paying fees for Weiss's lectures. He asserted that his administration "considers Professor Weiss as a teacher employed at the *Berg-Eleven-Institut*, for which he receives, like all other teachers, like the *Ober-Medicinal* Klaproth, Professor Fischer etc., the admittedly relatively high salary of 1,000 R[eichsthalers] from the main fund for the mining students [*Haupt-Eleven-Kasse*]."[25] In December 1810, the minister of the interior declared that he agreed to the regulation concerning the salaries of Klaproth, Fischer, and Weiss, which also implied that there was no longer an independent lecture series on the mining sciences. The mineralogical cabinet remained at its original place at the Royal mint until summer of 1815, when it was moved to the university building.

The political negotiations between the Sections for Mining Affairs and for Public Education make crystal clear that the young Berlin university was not exclusively a place of pure, disinterested science and neohumanist *Bildung*. The negotiations also took into account the mining administration's interests in the teaching of useful knowledge. But with the integration of the scientific part of the lectures on mining science into the university curriculum, one former aim of the mining administration was abandoned: the teaching of the technological part of mining science. The university could neither give the future mining officials instruction in drawing, nor impart knowledge concerning the construction of mines, mechanical engineering, surveying, metallurgical assaying, and smelting techniques. As a result, in 1818 the mining administration complained that "the teaching of the *Eleven* in assaying and metallurgy" was not covered by the university. It briefly considered hiring its own teachers and setting up a new laboratory, but this idea was quickly abandoned because the Ministry of the Interior gave assurances that it would "take care to allow for this instruction in the filling of the professorship of chemistry at the university here [Berlin] and the Academy of Sciences."[26] At the time, the university was searching for Klaproth's successor. However, in February 1822, in the appointment of Eilhard Mitscherlich (1794–1863), the Ministry's pledge was not upheld. Mitscherlich did have interests in the useful sciences—he was also a teacher at the War School, the United Artillery and Engineering School (Vereinigte Artillerie- und Ingenieurschule), and the Friedrich-Wilhelms-Institut for army surgeons—but, as far as we know, he did not maintain relationships with the mining administration.

TECHNOLOGY, AGRICULTURAL SCIENCE, AND FOREST SCIENCE

In September 1810, the university hired yet another leading state official: Sigismund Friedrich Hermbstaedt, who had long been a member of the Department of Trade and Industry. Hermbstaedt was first appointed *professor extraordinarius* and in November 1811 received a regular full professorship for "chemistry and technology." He taught "general experimental chemistry" and "experimental physics" four to six hours a week, as well as a daily course on "general technology" (*Allgemeine Technologie*) in the summer semester. Like his role model, the Göttingen professor Johann Beckmann, he was a proponent of a broad discipline of technology (*Technologie*) that encompassed knowledge about all branches of industry, including agriculture. Hence in the winter semester from 1811–1812 to 1825–1826 he also offered an additional six-hour lecture on "agronomical chemistry with application to agricultural industry." In 1829, lectures on "forest chemistry" (*Forstchemie*) and "forest technology" (*Forstechnologie*) followed. Alongside these lectures, he also taught a whole series of specialized technological topics including "chemistry in relation to dyeing along with experiments" (in summer semester 1811) and "commodity science [*Warenkunde*] based on my own notebooks" (in winter semester 1811–1812). After Klaproth's death in 1817, he also took over his chemistry courses, until in February 1822 Mitscherlich was named as Klaproth's successor.[27]

The University of Berlin not only established a chair for technology and took over the scientific education of the Prussian mining officials; it also created professorships for cameral and agricultural science as well as forest science. In August 1810, Albrecht Daniel Thaer was appointed a *professor extraordinarius* of cameral science. Until 1819, he taught two courses for future state officials, a course on the management of landed estates—"rural economy in connection with the so-called cameralistic and agricultural policy"—and a course on "agriculture and animal husbandry in their specific branches."[28] In addition, he combined these lectures with practical instruction at his estate in Möglin.

Thaer had studied medicine at the University of Göttingen and then practiced as a physician in his hometown of Celle. In 1784, he became a member of the Royal Agricultural Society of Celle, one of the numerous economic societies in the German states, and began to carry out agricultural experiments. A few years later, he established a model farm for performing agricultural experiments and demonstrating advanced farming technologies, which were adopted in part from England. In his publications he argued for a scientific training of farmers, and he soon became known as an authority in agricultural science. In spring of 1802, he founded an agricultural school at Celle, which organized public lectures on agricultural science, chemistry, and botany. When in the same year Karl August von Hardenberg paid Thaer a visit in Celle, he made him an offer to establish, with state support, a model farm along with an agricultural school in Prussia. In 1804 Thaer bought an estate in Möglin, near Bad Freienwalde in the Oderbruch region. Beginning in November 1806, he taught there a one-year course that combined agricultural practice with agricultural science and neighboring disciplines. The lectures took place in the morning and in winter on several afternoons, while in summer all afternoons were reserved for practical work. As a professor at the University of Berlin, Thaer lectured only in winter. His students had the choice between two ways of combining theory with practice. One was to first participate in his one-year course in Möglin, then attend his lectures at the university in winter semester, and afterward add a special course on the economy of agriculture, again in Möglin. In Thaer's view, this was the ideal way to educate and train a state official competent in various areas of agriculture. The alternative way consisted of a combination of his own university lectures with additional university lectures on chemistry, physics, botany, zoology, and mathematics, supplemented by practical work at his estate in the summer.

In spring of 1811, the university hired the Prussian forestry official Georg Ludwig Hartig (1764–1837) as a lecturer for forest science.[29] Hartig had first completed an apprenticeship in forestry with his father and uncle and, after a brief attendance at the University of Giessen, assumed a position as forester in Hungen (Wetterau), where he founded a school for forest science. As Prussia's main forester (*Oberlandesförster*), he was responsible for the whole of Prussian forests. After he had established a private "Academy of Forestry" near Berlin, his private students received tuition-free access to his university lectures, while the university's students were admitted to the practical instruction at his academy.

In 1819, at Hartig's instigation, the teaching of forest science was expanded with the appointment of the forester Friedrich Wilhelm Leopold Pfeil (1783–1859).[30] After his nomination as an honorary doctor and *professor extraordinarius* of the university, Pfeil founded the Institute of Forest Science (Forstwissenschaftliches Institut) at the University of Berlin. He used the institute to announce and coordinate the university lectures of relevance to forestry, comprising the lectures on chemistry, botany, zoology, physics, mathematics, technology, agricultural science, and forest science. Like Thaer and Hartig, Pfeil also enabled his students to gain practical experience in the field. The connection of theory and practice on the model of Freiberg Mining Academy was thereby continued at the young Berlin university.

All in all, compared to the eighteenth-century Prussian universities the University of Berlin was a place of relative academic freedom. Yet it was not an institution that would have fostered pure science and neohumanist *Bildung* at the expense of useful, technological knowledge. On the contrary, it continued to teach technological knowledge and useful sciences to future officials doing service in the technical departments of the Prussian state administration. Furthermore, a significant number of its professors were both natural scientists and men who possessed technological knowledge. Thus, they were also teachers at the War School, the United Artillery and Engineering School, the Friedrich-Wilhelms-Institut for army surgeons, as well as at the Academy of Civil Engineering and Architecture and the Industrial Institute, which will be discussed in the next chapters. As Norton Wise has observed, these men "acted as mediators who carried some of the goals of the technical schools into their own careers and into the training that they began to promote at the University."[31]

15 THE ACADEMY OF CIVIL ENGINEERING AND ARCHITECTURE

Since the foundation of the General Building Department in June 1770, the Prussian state administration had been seeking officials with technical expertise and practical experience in land surveying and in the construction of canals, streets, bridges, and other areas of civil and hydraulic engineering. In public building, the building councilors (*Bauräte*) were convinced, too much weight had been given to aesthetics and grand architecture (*Prachtbau*), while knowledge about civil engineering and the science of civil engineering (*Bauwissenschaft*) had long been neglected.[1] What Prussia needed, they argued, were fewer ingenious architects and more competent land surveyors and master builders. Consequently, in 1771 the minister responsible for education, Karl Abraham von Zedlitz (1731–1793), proposed the establishment of a school that would be set up "not for the so-called higher architecture, but rather for such construction operations as serve the general improvement of the country."[2]

While Zedlitz's proposal fell on deaf ears, his colleague Minister vom Hagen achieved a small success. As we have seen in chapter 9, in the spring of 1770 Hagen used the argument that there was a demand for expert building officials as a vehicle to get the king's approval for the establishment of a lecture series concerned with the useful sciences. Thus, from spring of 1771, the building councilor Friedrich Holsche taught mathematics, practical geometry, civic architecture (*Zivilbaukunst*), mechanics, and hydrostatics for future building officials in the context of this lecture series. In the fall of 1779, his lectures were canceled on the initiative on Minister von Heinitz, who wished to limit the spending of the mining administration to the training of its own officials.

After Prussia gained new territories through the first Partition of Poland in 1772, the General Building Department intensified its land surveying efforts, especially in the regions of Silesia, the formerly Polish Royal Prussia, and the Diocese of Ermland. It also organized new drainage projects in the marshlands of the Warthe, Oder, and Netze rivers.[3] Hence its demand for surveyors and civil engineers (*Baumeister*) increased. In 1776, a French mathematician by the name of Marsson gave lectures on civil engineering and architecture, with Zedlitz's ministerial backing, for both military engineers and building officials. The lectures were held at the Berlin castle under the name of *"École de genie et d'architecture."* Like the lectures for building officials organized by the Mining and Smelting Works Department, this school did not last long. A reorganization in 1788 preserved only the military part, which was moved to Potsdam to become a school for military engineers.[4]

REORGANIZATION OF THE ACADEMY OF ARTS?

In addition to proposals to found a new school for civil engineering and architecture, reform-minded ministers also explored possibilities for using an older institution for this purpose: the Royal Prussian Academy of Arts in Berlin. Founded in 1696 as the Academy of Painting, Sculpture, and Architecture and renamed in 1704 the Royal Prussian Academy of Arts and Mechanical Sciences, this institution had experienced a decline over the course of the eighteenth century. During the reign of Friedrich Wilhelm I (1713–1740), its architecture class had already sunk to the level of an ordinary school of drawing.[5] A few decades later, the Academy of Arts was apparently in a state of complete disorganization. When in the spring of 1770 Minister vom Hagen was looking for an appropriate location for the planned lecture series on the useful sciences and inspected the Academy of Arts, he concluded that it was "completely deteriorated."[6]

In the late 1780s, Minister von Heinitz, who since 1786 was also responsible for the Academy of Arts, set out to reorganize this institution.[7] Supported by the influential court building councilor (*Oberhofbaurat*) Friedrich Christian Becherer (1747–1823), he wished to turn it into an institution for the improvement of architecture as well as for the training of building officials and private master builders. Accordingly, in 1790 an "Architectural School at the Academy of Arts" (*Architektonische Lehranstalt bei der Akademie der Künste*) was founded. But this school, too, had only a brief existence of three to four years. Heinitz only succeeded in establishing small local art schools in Berlin, Königsberg, and several other Prussian cities.

After the Second and Third Partitions of Poland in 1793 and 1795, and the resultant territorial expansion of Prussia, the lack of expert officials in the General Building Department and the building administrations in the provinces became an acute problem. The new territory had to be measured, and the construction of new canals and settlements in the drained marsh-land regions and in the iron-smelting centers in Upper Silesia required civil engineering, for which the existing personnel clearly were insufficiently experienced and specialized.[8] Because the Architectural School at the Academy of Arts emphasized grand architecture, members of the General Building Department had long been skeptical of this institution. To meet the new demands, they looked instead for possibilities to foster expertise in surveying, hydraulic engineering, and practical, functional building. Thus, in 1793 David Gilly (1748–1808) and other members of the General Building Department founded a private school for the professional training of young men in architecture, civil engineering, and hydraulic engineering (*Wasserbaukunst*).[9] The school offered courses in land surveying, machine construction (mechanical engineering), hydrostatics, hydraulic engineering, architectonic drawing, building construction, and mathematics to future building officials. The teachers were members of the General Building Department, among them Gilly and the building councilor Johann Albert Eytelwein (1764–1848).

Gilly and Eytelwein embodied the figure of the scientific-technological expert in the field of civil engineering and the associated mathematical and natural sciences. Gilly had trained as a land surveyor and gained extensive practical experience in the drainage of the Netze and Warthe marshlands and as a building director in Pomerania. He was the author of numerous publications, many of which were published in the context of his efforts to establish a new school for the training of building officials. Eytelwein, a student of the mathematician and artillery officer Georg Friedrich L. von Tempelhoff, had trained to be a bombardier and ordnance technician in the Prussian artillery corps. After leaving military service in 1790, he became an inspector of the dikes in the Oderbruch region and later a councilor at the General Building Department. In 1803, he became a member of the Royal Prussian Academy of Sciences.[10]

Although it received some funding from the General Directory, Gilly's and Eytelwein's private school could only financially support itself until the winter of 1796–1797. Even so, officials in the General Building Department and the General Directory, including Minister von Heinitz, regarded the school as the right initiative; they thus searched for ways to establish a successor institution. When in 1797 the reform-minded Minister Friedrich Leopold von Schroetter (1743–1815) became responsible for the General Building Department, he endorsed these efforts, and in November 1797 Gilly and other building officials were commissioned to devise a concrete plan for a new school. Their finished plan foresaw a school with a five-year program that comprised "scientific courses" (*scientivische Collegien*) in winter and practical instruction in surveying, leveling, and civil engineering in summer. This was a combination of theory and practice comparable to the teaching at the mining academies. The plan also proposed a curriculum that further specified the teaching subjects. The General Directory approved the plan, and Minister von Schroetter subsequently submitted it to Friedrich Wilhelm III (reigned 1797–1840).

However, the king refused his consent. While he agreed that the Academy of Arts focused too strongly on grand architecture, and therefore was not the right place for the training of the state's professional surveyors and civil engineers, he also declared that he "eventually decided against [the foundation of] an entirely new institute." Arguing that the "foundations of grand architecture [*Prachtbau*] and economic architecture [*Oekonomie-Baukunst*] are the same," and that the Academy of Arts would be able to provide insight into these foundations, he added, "hence there is no need for new teachers with new salaries."[11] The objection of Friedrich Wilhelm III was in principle the same as that of Friedrich II when in 1770 the establishment of a mining academy had been under consideration: new schools were too expensive. Therefore, the king was in favor of yet another reorganization of the Academy of Arts. The latter was strongly supported by the court architect Carl Gotthard Langhans (1732–1808) and the court building councilor Becherer, who were skeptical toward the technical orientation of the General Building Department and instead argued for a stronger emphasis on "taste and art" (*Geschmack und Kunst*) in architecture.[12] In the following decades, the alternative between *Oekonomie-Baukunst* (civil engineering) and aesthetic architecture would be a predominant issue in debates about the Academy of Civil Engineering and Architecture.

THEORY AND PRACTICE

After further petitions, in March 1799 the king finally gave his approval for the establishment of an Academy of Civil Engineering and Architecture (Bauakademie). The result was a compromise, negotiated between members of the Academy of Arts, the General Building Department, and further persons influential at court such as Langhans and Becherer. What was called the Bauakademie was, on the one hand, not a truly new institution, but rather the Academy of Arts' renamed class of architecture. Accordingly, the memorandum of the foundation commission stated that the "already existing architectural school at the Academy of Arts" was to be "elevated to an educational establishment for general building instruction" and be given "the name of a Royal *Bauakademie*."[13] At the same time, however, the curriculum previously elaborated by members of the General Building Department was accepted, with only minor corrections, which placed civil engineering at the heart of the curriculum. Theory, that is, classroom teaching, was combined with practice, that is, surveying in the field and work at royal construction sites, which for practical reasons took place only in summer.[14] As Kathryn Olesko has pointed out, "The *Bauakademie's* curriculum thus had a strong emphasis upon the practical."[15] Furthermore, among the four directors of the Bauakademie, three were members of the General Building Department, namely Gilly, Eytelwein, and the mining councilor Heinrich August Riedel senior (1748–1810). Becherer, the representative of aesthetic architecture, was the fourth director as well as a teacher, while Langhans and the court sculptor Johann Gottfried Schadow (1764–1850) offered merely supplemental courses.

On April 21, 1799, the Bauakademie commenced instruction in the rooms of the Academy of Arts, with ten students (*Bau-Eleven*) to be trained as building officials. The admission requirements stipulated that the students should have reached the age of fifteen, those in the courses on hand drawing (*freies Handzeichnen*) excepted, which could be attended by students as young as twelve.[16] The average age of the students thus corresponded with the entrance age of apprentices in Prussia, which was between fourteen and fifteen. The students were required to have completed the middle level (*Secunda*) of a *Gymnasium* (upper secondary school), and thus have basic knowledge of French and Latin.

Hence, the Bauakademie was at first not regarded as an institution on the same level as the universities. Like the mining academies and the Industrial Institute (Gewerbeinstitut) founded some twenty years later, it was a new type of teaching institution that combined practical training with technological and scientific instruction related to civil engineering and architecture. Accordingly, in a report from February 1801 the three directors, Gilly, Eytelwein, and Riedel, pointed out that the Bauakademie aimed to educate civil engineers (*Baumeister*) for the state rather than mere artisans or, at the other extreme, professors of architecture. The teachers at the Bauakademie selected their subjects according to professional demands, and anyone who wished "to learn the higher sciences and arts" had to go to other institutions.[17] Like the future mining officials, after finishing their course of study the *Bau-Eleven* became *Cadets* and commenced professional work in the General Building Department or the administration in the provinces.

The regular course of study at the young Bauakademie lasted four years, divided into a one-and-a-half-year course concentrating on land surveying and leveling and a two-and-a-half-year course concerned with civil engineering. Successful completion of the course on land surveying was a formal requirement to attend the course on civil engineering.[18] Classroom instruction comprised various fields of mathematics, including practical mathematics (arithmetic, algebra, geometry, optics, and perspective), physical sciences (statics, hydrostatics, hydraulics, mechanics of solid bodies), and a variety of practical topics such as land surveying and leveling; rules of building; urban construction; river rectificaton; the building of dikes, sluices, ports, and bridges (*Wasserbaukunst*); the construction of streets; and the construction of machines. Instruction in drawing was extensive and included freehand drawing, architectural drawing, the drawing of site plans as well as machine drawing. The lectures further extended to management (*Geschäftsstil*) and the history of the art of building.

The practical part of the courses ranged from land surveying and leveling to work at the state's building sites, first in Berlin and then in the provinces. While practical work in Berlin took place on two days, with the remaining four days of the week being reserved for classroom teaching, in the provinces "practice" meant more extended, ordinary work on site. Administrations in the provinces were ordered to provide information about their building projects and to create opportunities for advanced students to participate in these projects, for example as supervisors or managers of building materials.[19]

With its combination of theory in the form of classroom instruction and practice in the form of work in the field and at construction sites, the young Bauakademie followed the model of dual theoretical education and practical training previously established at the mining academies in Freiberg and Schemnitz. This dual model clearly differed from the postrevolutionary French system, in which theoretical instruction at the École Polytechnique came first, followed by training at the more practically oriented, specialized *Écoles d'application*.[20] Because the combination of classroom teaching and practical exercise at a single institution was also organized elsewhere in the German states—at agricultural schools (such as Albrecht Thaer's in Möglin), schools of forestry (such as Georg Ludwig Hartig's near Berlin), chemical-pharmaceutical boarding schools (such as Hermbstaedt's in Berlin), the eighteenth-century artillery schools, Scharnhorst's War School, the United Artillery and Engineering School, the Industrial Institute, and even at the newly founded University of Berlin with respect to the teaching of agriculture and forestry—it is very likely that the historical actors in Prussia regarded this model as more successful, or more promising, than that of the French.[21]

TEACHING AND RESEARCH

The majority of the teachers at the Bauakademie were members of the General Building Department.[22] Since the foundation of the General Building Department in 1770, its members had actively contributed to the literature on the science of civil engineering (*Bauwissenschaft*). To their recurrent topics belonged hydraulic engineering and new, wood-saving building materials and techniques. As has been mentioned in several previous chapters, the shortage of wood stimulated research in forestry and in the field of civil engineering and architecture. Building officials investigated wood-saving building techniques as well as new building materials, along with new techniques of solid building (*Massivbau*).[23] As we have seen in chapter 1, some decades earlier even members of the Royal Prussian Academy of Sciences undertook similar investigations.

Eytelwein and Gilly were two of the most productive authors and teachers among the building officials, and they intensified their inquiry in the context of their teaching activities. Eytelwein published several textbooks on mechanics, applied mathematics, and especially hydraulics, among them a well-received *Handbook of Mechanics of Solid Bodies and Hydraulics* (*Handbuch der Mechanik fester Körper und der Hydraulik*), published in 1801.[24] Beginning in 1797, Gilly coedited, with other building officials, a professional journal entitled *Collection of Useful Essays and News Concerning the Art of Building* (*Sammlungen nützlicher Aufsätze und Nachrichten die Baukunst betreffend*).[25] In numerous articles, the authors argued for the introduction of professional training for building officials in the framework of a special technological and scientific school. In 1797–1798, Gilly published a two-volume handbook on *Landbaukunst* which also served as a guideline for teaching.[26] In an advertisement, he wrote that the construction of domestic and farm buildings relies "on theoretical and practical principles as well as on knowledge about the quality and preparation of building materials and the techniques of workers and builders, that is, on rules that one must never ignore."[27] The book contained numerous tables and precise drawings, which Gilly regarded as indispensable means to represent and transfer knowledge about building materials and techniques. It stimulated further investigations of these issues, which led to the publication of a plethora of brochures and booklets written specifically for use in teaching. As the historian Reinhart Strecke has pointed out, these publications were the result of innovative investigations, even though in hindsight some of them contained "wrong theoretical presuppositions" or offered unfeasible solutions to technical problems.[28] Whether these investigations and the teaching connected with them immediately led to success and actual improvements is of less interest than the fact that new institutions were established that organized instruction in building and at the same time sustained techno-logical and scientific research.

GRAND ARCHITECTURE OR CIVIL ENGINEERING?

In October 1800, the Bauakademie moved into the new mint on Werder-scher Markt, while the Academy of Arts remained in its location in the Marstall on Boulevard Unter den Linden. The new mint, erected from 1798 to 1800, was originally planned to accommodate the mint, the Mining and Smelting Works Department, and the royal mineralogical cabinet, but now also housed the General Building Department and the Bauakademie. This strengthened the bonds between the latter two institutions. Until the beginning of the anti-Napoleonic war in 1806, the number of students of the Bauakademie increased in the summer to ca. fifty and in winter, when the practical courses were not in session, to ca. one hundred. However, in the aftermath of the war and the reorganization of the Prussian state administration in 1808, and in the context of debates about the foundation of a university in Berlin, in 1809–1810 the Bauakademie was reunited with the Academy of Arts.[29] Alternative, money-saving proposals, such as trans-fer of the Bauakademie's scientific teaching to the University of Berlin, or the unification of the Academy of Arts with the new university, were even-tually rejected.[30] The directors of the Academy of Arts, the painter Johann Christoph Frisch (1738–1815) and, from 1816, the sculptor Johann Gott-fried Schadow, became directors of the Bauakademie as well. Consequently, the number of lectures on grand architecture increased, while the lectures on mathematics, natural sciences, and civil engineering were reduced.[31]

In the years to come, the controversy continued between the members of the Academy of Arts and the members of the General Building Department (Oberbaudeputation) concerning the teaching of civil engineering and of grand architecture.[32] In 1817, the Minister of Culture, Karl Freiherr vom Stein zum Altenstein (1770–1840), who was responsible for the Academy of Arts as well as the Bauakademie, observed that the latter was "in decline." He instructed the mathematician Johann Georg Tralles (1763–1822), a professor at the University of Berlin, to elaborate a "scientific plan" for its renewal. In connection with this plan, Tralles argued that a professional training of "architects and technical experts [*Techniker*]" was possible only at a "mathematical-technical teaching institution [*Mathematisch-Technische Lehranstalt*]" and that the Bauakademie had to become such an institution.

The minister of trade, who was responsible for the latter, received the plan favorably, and asked Eytelwein, since 1810 director of the Building Department, to comment on it. In his response, Eytelwein pointed out that "the lack of knowledgeable civil engineers [*Baumeister*] has proven to be a disadvantage for the building administration," and that "improvements in the deteriorated state of the Bauakademie" by means of its reorganization should be made.[33]

In April 1824, after its transfer to the competence of the Ministry of Trade, the Bauakademie was separated again from the Academy of Arts.[34] Shortly afterward, Eytelwein was appointed its director. From then on, the teaching of aesthetics and grand architecture was entirely in the responsibility of the Academy of Arts, while the Bauakademie taught exclusively civil engineering and related sciences. The previous division of teaching into a course for the training of surveyors followed by a course for the training of civil engineers was reintroduced, as was the combination of classroom teaching and practical training. The technological part of the curriculum regained the same broad range that it had before 1809–1810, and the scientific lectures included more courses on higher mathematics. Friedrich Christian Accum (1769–1838), who was also a professor at the Industrial Institute (see chapter 15), taught additional courses on physics, chemistry, and mineralogy "in relation to building." The specification "in relation to building" highlights Eytelwein's efforts to couple the natural sciences with civil engineering.[35] In 1830 Eytelwein was forced to retire on account of poor health. Under his directorship the number of students had climbed steadily to approximately 140.[36]

SPECIALIZATION: SURVEYORS, MASTER BUILDERS, BUILDING INSPECTORS

In 1831, the influential official Christian Peter Wilhelm Beuth (1781–1853) became the director of the Bauakademie, now renamed the Allgemeine Bauschule.[37] The Bauakademie moved into a new building on Werderscher Markt, erected between 1832 and 1836 and designed by Beuth's friend, the famous Prussian architect Karl Friedrich Schinkel (1781–1841). In 1830, Schinkel had become the director of the General Building Department, which implied that he also gained influence on the Bauakademie, although he was never one of its teachers.

Under Beuth's directorship from 1831 until 1845, the Bauakademie extended the duration of practical training and continued the previous teaching of civil engineering and related sciences, while also leaving room for

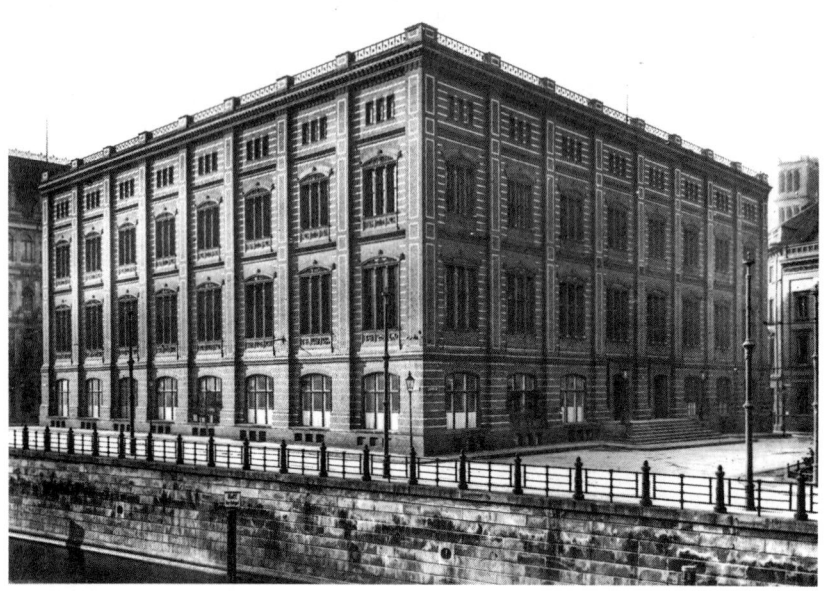

FIGURE 15.1
The Bauakademie, erected 1832—1836 after a design by Karl Friedrich Schinkel, illustration from 1905.
Source: Landesarchiv Berlin, figure no. II. 3413.

lectures on grand architecture. Beuth, who was also the director of the Industrial Institute, divided the course of study into three successive courses. Each course was completed by examination, and led to different professional specializations and ranks in the administrative hierarchy. First came the training of surveyors, who were now required to do at least one year of practical work before they were admitted to the exam. Completion of this predominantly practical phase of training was required for all building officials. The second course comprised the education and training of civil engineers named master builders (*Baumeister*), and was subdivided into a two-year-long theoretical part followed by a practical part of the same duration. This prolongation of practical training was the main reason for the considerable extension of the education and training of master builders from a course of two and a half years at the young Bauakademie to a four-year course under Beuth's directorship. The third and final course of one year, to which only the best students were admitted, offered training in architecture or in building technology and mathematics to qualify as a "building inspector" (*Bauinspektor*), the highest professional rank among the state's civil engineers.[38]

The classroom teaching for the master builders comprised a broad range of technological and scientific disciplines, and it put a strong emphasis on advanced mathematics. Concerning the natural sciences, it included mechanics and statics of solid bodies, hydrodynamics, "physics," chemistry, mineralogy, and, for the first time, aerodynamics and botany. The teaching at the Bauakademie and at the Industrial Institute intersected with respect to the latter, as four teachers—Friedrich Christian Accum, Adolf Ferdinand Wenceslaus Brix (1798–1870), Peter Gustav Lejeune Dirichlet (1805–1859), and Ernst Ludwig Schubarth (1797–1868)—also taught at the Industrial Institute and even offered their Bauakademie courses at the latter. Hence, the Bauakademie and the Industrial Institute became relatives, even though Beuth did not allow higher courses in mathematics.[39] The two schools also differed with respect to the future occupations of their students. While the Industrial Institute trained engineers for private industry (see chapter 15), the Bauakademie primarily trained officials.[40]

The building inspectors specialized either in hydraulic engineering (*Wasserbau*) and construction of machines (*Maschinenbau*; mechanical engineering) or in urban construction (*Stadtbau*), which also included grand architecture (*Prachtbau*). Their education and training concentrated more narrowly on subjects attuned to engineering or architecture, respectively, but they were also instructed in mathematics (advanced analysis and theory of curves), analytical dynamics, and the history of the art of building.[41] The young architects Friedrich August Stüler (1800–1865) and Wilhelm Stier (1799–1856) taught grand architecture, aesthetic design, and art historical courses, while Gotthilf Heinrich Ludwig Hagen (1797–1884), who was also a member of the General Building Department and (from 1842) of the Royal Prussian Academy of Sciences, was responsible for the teaching of hydraulic engineering. In this way, Beuth attempted to reconcile the two opposed camps, emphasizing in the two specializations either the aesthetics of architecture or civil engineering and related sciences. In the General Building Department, however, the highest positions, that of councilors, were reserved for those building inspectors who had completed courses in engineering.

After Beuth retired, the old controversy about the relationship of grand architecture to civil engineering was reignited. In 1846, the architect Stüler complained in a memorandum that the Bauakademie had "not paid sufficient attention to the training of artistic skills and had subordinated the former to [the students'] scientific and theoretical achievement."[42] His view was supported by the Society of Architects (Architektenverein), founded in 1824, which proposed the elevation of the Bauakademie to a level corresponding to that of a university. Accordingly, they also proposed to remove the training of land surveyors from the Bauakademie's remit.

When in the aftermath of the revolution in 1848 the Bauakademie was reorganized, its admission rules stipulated for the first time that its students must have completed the *Gymnasium*.[43] However, land surveying and leveling remained part of the curriculum. Likewise, in the following decades the combination of technological and scientific knowledge, and of theory and practice, was preserved. Further specializations were introduced as well as new disciplines—most importantly railroad construction and geology. The number of students increased significantly from 201 in 1849 to 547 in 1859–1860 and 655 in 1869–1870. The number of students fluctuated between 800 and 1,000 in the 1870s, that is, in the years running up to its unification in 1879 with the Industrial Institute to form the Royal Technical University of Berlin (Königlich Technische Hochschule zu Berlin).

Like the mining academies at Freiberg and Schemnitz, the Bauakademie was a new type of school established for the education and training of expert officials. It taught the natural and mathematical sciences as well as various technological disciplines related to civil engineering and architecture. Further, it combined classroom teaching with practical work at the state's construction sites and surveying in the field. The Bauakademie was Prussia's first technological and scientific institution that organized useful science in a sustained way, even though it was not an uncontested institution.

16 A NEW INDUSTRIAL POLICY: THE INDUSTRIAL INSTITUTE

Beginning in the 1770s, the Department of Mining and Smelting Works and the General Building Department organized education and training in science and technology to qualify men for state service. The Department of Trade and Industry in the General Directory undertook similar efforts, but these were aimed at a different group of students: at first the sons of craftsmen and factory owners, and later future engineers. While the Prussian state was directly responsible for the management of mining and metal production as well as for land surveying, hydraulic engineering, road construction, railways, and public buildings, and thus needed competent officials who could deal with these subjects, there was also a large private industrial sector whose improvement depended significantly on the initiative of entrepreneurs. But until the mid-nineteenth century, the state fostered private industry as well. Its activities ranged from various kinds of protection, financial support, the organization of industrial espionage, and donations of machines all the way to the education and training of technical experts and the organization of experiments and technological projects. For example, from the fall of 1796 until the spring of 1799, Minister Carl August von Struensee, head of the Department of Trade and Industry, organized a project that investigated the production of sugar using maple syrup. The maple sugar project competed with Franz Carl Achard's sugar beet project that was supported by Minister von Heinitz (see chapter 1). It was carried out by the apothecary and chemist Sigismund Friedrich Hermbstaedt and Gottlob Johann Christian Kunth (1757–1829), a leading official and expert in Struensee's department and the main private teacher of Alexander and Wilhelm von Humboldt from 1777 until 1789.[1] In 1796 Kunth was named head of a small circle of experts within the Department of Trade and Industry, the Technical Deputation (*Technische Deputation*), which became an important instrument in the promotion of mechanization of Prussian industry.

Beginning in 1801, Hermbstaedt and Kunth organized public chemical lectures for Berlin's dyers, calico printers, and textile fabricants. Hermbstaedt is yet another example of the figure of scientific-technological expert in Prussia. Originally trained as an apothecary, he was soon recognized for his achievements as a chemist. In 1796 he entered civil service in the Department of Trade and Industry, and a year afterward he became a member of the department's newly founded Technical Deputation. The latter position enabled him to continue his collaboration with Kunth and extend it to the organization of public chemical teaching. In 1802, he moved into a new building, purpose-built for him, the "*Chemicus* of the Factory Department."[2] The building comprised a lecture room and a large laboratory on the ground floor, as well as storage rooms in the cellar. On the first floor, there was a room for the physical and chemical instruments, a room for the mineralogical and the technological collection, a library, a private lodging for Hermbstaedt, and three small rooms for Hermbstaedt's assistants.

In the years that followed, Hermbstaedt and Kunth were major players in the state's effort to promote industry and technological-scientific knowledge relevant for industrial innovation. Kunth was the major organizer of Prussian industrial espionage in England, and he also elaborated numerous pedagogical proposals for the improvement of scientific education. In February 1816, at the age of fifty-nine, he retired, while Hermbstaedt remained an influential figure in the Technical Deputation, who was active until shortly before his death in 1833.

THE TECHNICAL DEPUTATION

The Technical Deputation, headed by Kunth between 1796 and 1809, was an advisory commission of expert officials and entrepreneurs. Its function can be best understood when we compare it to Minister von Heinitz's circle of experts, although it had a more formal status in the Prussian bureaucracy than Heinitz's circle ever had.[3] We have seen that in his circle, Heinitz brought together scientific-technological experts, including men such as Klaproth, Gerhard, Humboldt, and Karsten. These men carried out experiments, supervised the experiments of technical experts, and aided Heinitz in the organization of the education and training of officials responsible for mining, civil engineering, and the Royal Prussian Porcelain Manufactory. In addition, Heinitz recruited a number of lower technical officials charged with the transfer of knowledge from abroad and the introduction of steam engines in Prussian mining and at the Royal Prussian Porcelain Manufactory. The Technical Deputation was a similar cadre of experts, and its basic function mirrored that of Heinitz's circle: the acquisition and dissemination of useful knowledge.

Thus, a state order from December 1808 stipulated that the Technical Deputation should assemble men who had "the scientific or practical education [*praktische Bildung*]" necessary "to follow the scientific aspects of the discipline of technology [*Gewerbekunde*] and its progress."[4] In other words, the order highlighted the need for a group of experts who combined scientific and practical knowledge and had insight into *Gewerbekunde*, that is, "technology" in Beckmann's sense of a discipline that collects knowledge about industry.[5] This would be the basis for competent judgments and the distribution of useful knowledge to entrepreneurs. Another memorandum from June 1812 described the main task of the Technical Deputation as "the reading of relevant journals and other publications, and observation of what is going on in the large field of industry [*Gewerbsamkeit*]."[6] These descriptions did not, however, fully disclose the group's methods of acquiring useful knowledge.

The Technical Deputation was also an agency of state-organized espionage and semilegal transfer of know-how and machines from England and other foreign countries. These kinds of activities started even before the Technical Deputation was founded. In 1787, when Karl Freiherr vom Stein (1757–1831) wanted to visit mines and factories in Cornwall, the British authorities were convinced that he came "with the patriotic intention to kidnap away what useful knowledge and men he may find in his way."[7] In the following decades, Kunth organized a veritable system of Prussian spies in London and Paris, who corresponded with the Berlin authorities in secret codes and sent them material under cover of shipments of physical instruments. Kunth also organized travels of members of the Technical Deputation to companies in other countries, among them the Cockerill company in the Netherlands, famous for its machinery construction. This paved the way for the company's establishment in Berlin in 1815. The new branch, settled with help of the state in a building located on Friedrichstraße, produced machines for wool fabrication.[8]

BEUTH'S SYSTEM

The reforms initiated after Prussia's defeat by the Napoleonic troops at Jena and Auerstedt led to the dissolution of the Technical Deputation in 1809 and its reestablishment in October 1811. In July 1819, Christian Peter Wilhelm Beuth became the director of the new Technical Deputation, which, apart from Beuth, included the following members: Sigismund Friedrich Hermbstaedt, the factory inspectors (*Fabrikenkommissare*) Johann Gottfried

May, Johann Heinrich Weber, and Georg Anton Frank, senior mining councilor Schaffrinsky, senior building councilor August Leopold Crelle, factory
commissioner Severin, and senior building councilor Karl Friedrich Schinkel.[9] Hence, from 1819 onward all members of the Technical Deputation
were officials.

Beuth was a highly influential official who played a leading role in the
mechanization of Prussian private industry. Although he was a commoner
and thus never had the formal status of a minister, his power is comparable
only to that of Minister von Heinitz in the late eighteenth century. From
1798 to 1801, Beuth had studied cameral science and law at the University
of Halle and then entered Prussian state service at the age of twenty. After
years of service in diverse areas of administration, interrupted for one year
during the anti-Napoleonic War, he entered the Department of Trade and
Industry, headed by Kunth, in 1814. The following years of collaboration
with Kunth and Hermbstaedt were formative for his policy. He began to
engage in debates about the improvement of industry and technical expertise. After his appointment as director of the Technical Deputation, he
steadily climbed the career ladder to a position that almost equaled that of
a minister. In January 1828, he was named head of the Interior Ministry's
Section for Trade and Industry, and in 1831 director of the Academy of Civil
Engineering and Architecture (Bauakademie).[10] He retained these positions
until he retired in 1845, at the age of sixty-four.[11]

Beuth extended his technological knowledge through visits to numerous industrial establishments in Britain and other countries. In the summer
of 1826, he traveled to Britain for the second time and in the company of
his friend and fellow member of the Technical Deputation, Karl Friedrich
Schinkel. The two men visited, among other things, the engineering works
and plants of the Gas Light and Coke Company in London, the Manchester cotton factories, and Talbot's chemical factory at Glasgow. Beuth also
organized countless technological travels of his officials and consultants.
For example, in 1819 he sent the young mechanic Franz Anton Egells (1788–
1854) to England in order to study machines. Egells, who had already gathered some knowledge about steam engines, visited the factory of Boulton
& Watt as well as engineering works in Manchester, Leeds, Sheffield, and
Birmingham. Back in Berlin, he received a patent for a technical innovation
of steam engines. In 1822, he founded a machine and engineering works
(*Maschinenbau-Anstalt*) in Berlin, with state support, where he began to construct steam engines. This was followed by the foundation of the "New Iron
Foundry of Berlin" in 1826, where August Borsig (1804–1854), a future lead-
ing Prussian entrepreneur who had just completed his studies at Beuth's

Industrial Institute, got a managerial position. By 1830, Egells's company employed several hundred workers and had established the first training workshop for machine construction in the German states.[12]

Travels organized by the Technical Deputation served not only the transfer of technological knowledge but also that of machines. On his personal travels to Britain in 1823 and 1826, Beuth acquired a number of machines for use in the textile industry. Based on the system of industrial espionage established by Kunth, he took measures to circumvent the British ban on the export of machines.[13] The machines transferred from England were examined and tested by the Technical Deputation, then replicated on the basis of precise drawings, and later donated to entrepreneurs. In order to distribute machines in Prussia, Beuth devised another sophisticated method that used his own students as distribution agents. Having negotiated with a factory owner about the employment of a former student, he organized the donation of state-paid machines to this student who then brought the machines with him to the factory. In the ideal case, the young engineer became a factory co-owner who invested his machines as well as his knowledge in the company.[14] Beuth's system guaranteed both employment for engineers trained at his institute and the proliferation of machines in Prussia.

BEUTH'S CIRCLE OF EXPERTS

In addition to reorganizing the Technical Deputation in 1819, Beuth founded a new learned society, the Association for the Advancement of Industry in Prussia (Verein zur Beförderung des Gewerbefleißes in Preußen), which received governmental approval in October 1820. Furthermore, in April 1821, he organized the foundation of a new school, first named the Technical School (Technische Schule) and from 1827 the Industrial Institute (Gewerbeinstitut). Beuth established a division of functions between the Technical Deputation, the Association, and the Industrial Institute, which entailed the transfer of several original functions of the Technical Deputation to the Industrial Institute and to the Association. In Beuth's view, the successful promotion of Prussian industry depended upon the three institutions all working toward to this common goal, with each performing its distinct, specialized functions, like the different organs of one organism.

Thus, teaching, which Kunth had defined vaguely as "public dissemination" of knowledge through the Technical Deputation, became formalized and the main function of the Industrial Institute.[15] Several members of the Technical Deputation became salaried teachers at the school. By contrast, the goal of the Association was the more informal dissemination of knowledge to the broader public, mainly through the annual publication of a journal, the *Verhandlungen des Vereins*, and exchange of knowledge and ideas at the Association's monthly meetings. The Association brought together a large number of officials, members of the military, men of science (such as Alexander and Wilhelm von Humboldt), artists and architects (such as Schadow and Schinkel), factory owners (such as Friedrich Krupp), merchants, and craftsmen. As its head, Beuth, declared at the first meeting, the Association wanted to bring together practical industrial "men dedicating themselves wholeheartedly to those sciences that are fundamental for industry [*Gewerbe*]."[16] The number of members increased from 367 in the year of the Association's foundation—with 110 officials being the largest subgroup—to an average of 900 between 1835 and 1845.[17]

EXPERIMENTS IN THE NEW BUILDING

Regarding the Technical Deputation, Beuth had elaborated three plans for its reorganization between June 1817 and January 1818. He pointed out that its main task was active acquisition of new knowledge relevant for industry, and that this included not only travels and knowledge transfer from abroad, but also experimentation and investigative technological projects.[18] Further, Beuth pointed out that in order to function as an investigative organization it was crucial that the Technical Deputation have access to laboratories and workshops. In 1821, he took a big step toward his goals through the acquisition of a representative house at Klosterstraße 36. A year afterward, he reported to Trade Minister Hans von Bülow (1774–1825) that the goal of the Technical Deputation was the collection and dissemination of "scientific and practical-technical knowledge." Further, he observed that the new building, which was designated the "Official Building [*Dienstlokal*] of the Technical Deputation," now offered sufficient space to perform even "large-scale, decisive experiments."[19]

The Technical Deputation's building comprised laboratories, a workshop, a collection of machines, a collection of models of machines, a collection of products and fabricated articles (*Produkten- und Fabrikatensammlung*), an engraving workshop, a library, and a meeting room for the Technical Deputation. In addition, there were two classrooms for the Technical School and a meeting room for the Association. 1829 saw the completion of an extension of the building, designed by Schinkel, for use primarily by the newly renamed Industrial Institute (Gewerbeinstitut) (figure 16.1).

For its chemical experiments, the Technical Deputation had altogether three laboratories at its disposal, with adjacent rooms for storing chemical

FIGURE 16.1
The building of the Technical Deputation and the Industrial Institute, with
Schinkel's extension in the foreground, painting by Eduard Gaertner (1830).
Source: Geheimes Staatsarchiv Preußischer Kulturbesitz 2014, 73.

and physical apparatus. There was a small laboratory containing a broad
range of instruments, vessels, and materials for use in a variety of different small-scale experiments. Two larger laboratories with vaulted ceilings
and fireproof chimneys were designed for carrying out large-scale experiments on the fabrication of glass, different kinds of steel, and so on. These
experiments are reminiscent of the chemical-technological experiments
Carl Abraham Gerhard had carried out in the laboratory of the mining
administration.[20]

INVESTIGATIVE ENGINEERING WORK

In addition to chemical experiments, Beuth also wanted the Technical Deputation to undertake investigative engineering work. The latter included
engineering design and the testing of machines installed temporarily in the
"machine collection." These machines were either imported from abroad,
mostly from England, or had been constructed in workshops in Berlin,
based on drawings transferred from abroad. Because England prohibited

the export of machines, machines were disassembled into parts in England and reconstructed in the rooms of the Technical Deputation. As Beuth explained in a letter to Austrian authorities in 1839, the members of the Technical Deputation then "carried out experiments concerning their utility and advantages." If the experiments were successful, they made precise technical drawings or scale models to be deposited in the collections of the Deputation. The machines themselves were transferred to master craftsmen (*Werkmeister*) with the request "to build replicas in appropriate number, taking into account the improvements discovered in the experiments."[21] The whole process was financed by the state, which donated the replicas, after their exhibition in the Technical Deputation's machine collection, to factory owners.

The second site of engineering work was the "model workshop." In Beuth's words, it was a place for "constructing models of the newest and best machines on the basis of a plan."[22] The models were scale models of machines for use in the textile industry, constructed from the same materials as the corresponding real machines. Both the model workshop and the laboratories of the Technical Deputation were also used for the Industrial Institute's practical courses.

How much experimental research and testing of machines the members of the Technical Deputation actually carried out in the decades under Beuth's directorship is an open question. Reality may have deviated from Beuth's plan, given the fact that the Technical Deputation had to evaluate an increasing number of patent applications.[23] Even so, what should be noted here is the similarity of the activities of the different state departments concerned with mining, building (civil engineering), and private industry. All three departments agreed that knowledge was a crucial factor for the promotion of Prussia's industry. They also agreed that industrially useful knowledge comprised scientific, technological, and practical-technical knowledge, and that the state, in addition to teaching and transfer of knowledge from abroad, also had to organize experimental research and technological investigation. The alliance between reform-minded officials, men of science, and technical experts forged after the Seven Years' War in order to foster innovation in state-managed industry did not disintegrate in a period with a stronger predilection for liberalism than for mercantilism.[24] On the contrary, it continued well into the nineteenth century, as did related efforts to produce new bodies of useful knowledge. As we will see next, the combination of theory and practice was also a hallmark of the Industrial Institute.[25]

THE INDUSTRIAL INSTITUTE

The Industrial Institute of Berlin was founded in 1821 as a higher school for the sons of craftsmen, factory owners, and merchants to improve knowledge in mathematics, physics, chemistry, and technical drawing, with an emphasis on the last-mentioned; of the twenty-four hours of teaching, from Monday to Saturday, twelve were dedicated to drawing.[26] Beuth wanted the boys to enter the school at an average age of fourteen, after they had completed primary school and before they would traditionally start an apprenticeship in a workshop.[27] The state administration selected a maximum of thirty students per year and also provided a stipend. The majority of students were to take a one-year course and then leave the school with or without a certificate of completion. Students who wanted to become builders, in particular, left after the first course to continue with a traditional apprenticeship.[28] Similar arrangements were also made for the establishment of one-year industrial schools (*Gewerbeschulen*) in the Prussian provinces.

A unique feature of the Industrial Institute in Berlin was that it offered an additional one-year upper course for the most talented students.[29] According to Beuth's plan, the upper course of the Industrial Institute comprised, in principle, the same disciplines as the lower one, that is, technical drawing, mathematical sciences, physics, and chemistry. But there were also important differences. First, students learned experimentation, the best students being invited to participate in the experiments of the Technical Deputation. Second, the teaching included a significant number of technological topics. What Beuth designated as "mathematical sciences" included algebra, arithmetic, geometry, trigonometry, statics, and mechanics as well as mechanical engineering ("practical knowledge about machines without proofs") and parts of technology (*Technologie*), including knowledge about goods produced with machines (*Warenkunde*). In the same vein, the natural sciences included knowledge about "products that are of interest to the technologist [*Technologe*]." The teaching of chemistry ranged from "theoretical chemistry" all the way to knowledge about useful materials, costs of chemical preparations, and uses of chemically prepared goods (*Warenkunde chemischer Fabrikate*). In the chemistry course students also received "instructions for practical work in the laboratories." Compared to the University of Berlin and the Bauakademie in the 1820s, the teaching at the young Industrial Institute was on a lower level. Hence, in 1822 Beuth pointed out that anyone who wished to learn more ought to attend either the university "or the Bauakademie with respect to the mathematical sciences and the various areas of technology [*Technik*]."[30]

PRACTICAL INSTRUCTION

As Peter Lundgreen and Wolfgang König have observed, the Industrial Institute of Berlin was an extraordinarily successful institution. Most of its students attained leading positions in the growing Prussian industrial sector.[31] This was due not least to the intensive practical training the institute offered alongside mathematical and scientific education. Practical instruction, or, as Beuth called it, "practical work," came after completion of the lower one-year course, and it took place in the model workshop and the laboratories of the Technical Deputation.[32] In the model workshop, students trained at a steam engine, in carpentry, and in metal working such as forging, filing, and casting. This type of practical work qualified them for metallurgical and mechanical professions and, in particular, for the new positions of the industrial age such as machine designer, machine builder (*Maschinenbauer, Werkmeister für Maschinenbau*), mechanical engineer (*Maschinenmeister*), and overseer in mechanized workshops and factories (*Werkstattvorsteher*).[33] Chemical training in the institute's laboratories prepared the way for positions as chemists, colorists, and advanced bleachers in chemical factories, calico-printing shops, and the dye industry. In addition, Beuth also organized practical training in factories. For example, beginning in 1825 the Royal Iron Foundry in Berlin offered the possibility of practical training. As such factory-based training could extend to longer periods, there was no clear boundary between a practical course and an apprenticeship training with a master. However, while it persisted in the building sector, the apprenticeship system in the 1820s was no longer an unquestioned tradition in the fields of mechanical engineering, metallurgy, and chemistry. With respect to professions in these fields, Beuth's method of practical training was quickly accepted.[34]

Hence, Beuth ensured that classroom learning, or "theory," was supplemented by "practice" in several ways: first, by practical work in the model workshop and the laboratories after the first year; second, by practical courses in factories; third, by chemical experimentation in the laboratories during the second year; fourth, by participation in the experiments of the Technical Deputation. The combination of theory and practice had been a major concern in the German discourse on the useful sciences. The mining academies of Freiberg and Schemnitz and the Bauakademie in Berlin systematically combined classroom instruction with practical training in mines, foundries, fields, and at construction sites. At the mining academies, in particular, classroom instruction, or theory, and practice were learned in parallel. Hence, Beuth followed a model of combining theory and practice already well established in Prussia and other German states. As pointed out in the previous chapter, this model differed significantly from the French system, which separated school-based "theoretical" teaching at the École Polytechnique from subsequent, more practically oriented teaching at the *Écoles d'application*. It also differed from the Conservatoire National des Arts et Métiers in Paris, founded in 1794, which was primarily a technical museum that offered additional teaching in the evening. Not always did the Prussians follow the example of the French, as some historians have supposed.[35]

FRIEDRICH ACCUM: SCIENTIFIC-TECHNOLOGICAL EXPERT

Most of the Industrial Institute's teachers were officials and members of the Technical Deputation. The first university-educated professor was Ernst Ludwig Schubarth (1797–1868), who had studied medicine at the University of Leipzig, received doctorates in medicine and philosophy at the University of Berlin, and was appointed a *professor extraordinarius* for materia medica and chemistry at Berlin in 1824. Schubarth was also a member of the Technical Deputation and a teacher at the Bauakademie.[36] He taught chemistry, physics, and technology until 1849. Today, he is known as the author of a textbook on technical chemistry, *Handbuch der technischen Chemie* (1839), which went through several editions. In 1822, the chemist Friedrich Christian Accum (1769–1838) was hired as a professor for chemistry and mineralogy with respect to their application to industry. Hermbstaedt, who also had the title of a professor, was not a regular teacher but occasionally offered lectures.[37]

Accum is perhaps the most interesting teacher at the Industrial Institute. Possessing practical skills and at the same time knowledgeable in the sciences and technology, Accum exemplifies the figure of scientific-technological expert at the Industrial Institute. After completion of a pharmaceutical apprenticeship in Hannover, his hometown, he moved to London in 1793 to take a position in a pharmacy. In London, he became more deeply involved in chemistry. In 1800, he established a laboratory and shop where he sold chemicals and carried out experiments. He also ran a boarding school for the teaching of pharmacy and chemistry, including chemical analysis. His shop soon became well known as a supplier for the growing community of chemists; he advertised that it "keeps constantly on sale in as pure a state as possible, all the Re-Agents and Articles of Research made use of in Experimental Chemistry, together with a complete Collection of Chemical Apparatus and Instruments."[38] Accum also lectured at other places: first, at the Royal Institution, where he was appointed "Assistant Chemical Operator" in 1801, and from 1803 at the Surrey Institution. In 1803, he published his first chemical textbook, *System of Theoretical and Practical Chemistry*, written in a popular style, followed by thirteen more books between 1804 and 1821—among them two books on the analysis of minerals, one on electrochemistry, one on gaslighting, and four on beverages and food. He also wrote numerous contributions to scientific journals. In 1815, he became a corresponding member of the Royal Prussian Academy of Sciences.

In addition to the preparation and sale of pure chemicals, Accum soon turned to another field of practice: coal gaslighting.[39] In June 1812, he became a member of the board of directors of the Gas Light and Coke Company, an incorporated joint-stock company that introduced a large urban gas-lighting network in London between 1812 and 1820. The directorship was the climax of his long association with the company. The idea to develop a network of pipes distributing gas to lamps came from Frederick Winsor, a German entrepreneur (originally named Friedrich Winzer) who had arrived in England in 1802. Winsor soon succeeded in forming a group of investors who were willing to support his project and promote the incorporation of a joint-stock company. The company was first called the National Light and Heat Company and renamed the Gas Light and Coke Company once it was incorporated in 1812. In fall of 1807, a committee of trustees, formed in the summer of the same year, organized the establishment of the first gasworks and large-scale demonstrations of gaslights in London. Shortly afterward, it invited Accum as a chemical expert with a public reputation to evaluate Winsor's apparatus and procedures and answer questions concerning the chemistry of gases. Accum gave evidence in support of Windsor's scheme and remained the company's chemical expert in the following years. As a director of the Gas Light and Coke Company he was involved in both managing the company and in day-to-day operations, including experimentation. In November 1813, he resigned from the position of director to become an independent expert in the field. Two years later, he published the first book on gaslighting, *A Practical Treatise on Gas-Light* (four editions, 1815–1818), whose translation into French and German led to attempts to introduce gaslighting in Paris, Vienna, and Freiburg.

In the time remaining until his move to Berlin in the winter of 1820–1821, Accum turned to yet another practical field: food adulteration. In 1820, he published a book on the topic, *A Treatise on Adulteration of Food and Culinary Poisons*, which received great public attention. In this book, he exposed common practices of toxic adulteration, such as the making of confectionery colored with copper salt or with vermilion, sweetening of wine with sugar of lead (lead acetate), and coloring of beer with oil of vitriol (sulfuric acid); he even listed individual firms responsible for adulteration.[40] A few months after the publication of this book, complaints about him were made to the Royal Institution, which forced him to leave England in the winter of 1820–1821. As we have seen in the previous chapter, on his return to Prussia Accum also became professor for chemistry and physics at the Bauakademie. In 1826, he published his last book, concerned with building materials, *Physische und Chemische Beschaffenheit der Baumaterialien*.

ENGINEERS AND *TECHNIKER*

Beuth had originally planned his school to be a teaching institution for boys, who would enter the school at the average age of fourteen and leave at the age of sixteen or seventeen. He had sought to train a broad range of artisans and craftsmen, including owners of small workshops. Over time, however, the character of the Industrial Institute changed from a middle-level school toward a higher academic institution for the education and training of engineers. The salaried engineer employed in a factory soon became the predominant position taken up by graduates of the school. In 1833, Beuth observed that there was a "constant demand from factory owners for good technical overseers [*Werkführer*]."[41] When two years later merchants from Königsberg wanted to employ an overseer for their steamship, they also appealed to Beuth. Beuth wrote back that many of his former students were employees in factories, as "their scientific and practical training enabled them to get easily familiar with the still unknown parts of a business." Another leading state official pointed out around the same time that the Industrial Institute no longer trained "ordinary craftsmen" but rather a small elite of men "who combined knowledge, talent, and practice."[42]

By the mid-nineteenth century, there were specific names for these men: *Techniker* (technical expert) and *Ingenieur* (engineer). *Ingenieur* then no longer had the narrow meaning of a military man concerned with demolition and fortification, but rather referred to all kinds of experts who had advanced mechanical knowledge and skill. *Techniker* was even more comprehensive, also including technical experts in chemical and metal works. The new terminology expressed technical experts' increasing awareness of the fact that they were a distinct social group and played a leading role in the development of industry and of the whole nation.[43] It was a long way from the itinerant inventor and technical expert of the sixteenth and seventeenth centuries, who offered his inventions and projects at courts, to the small group of expert officials in the eighteenth-century state administration, and thence to the self-confident social group of mid-nineteenth-century engineers and technical experts, educated and trained in the new type of advanced technological school.

In 1854, the engineer and director of the Saxon railways, Max Maria Freiherr von Weber (1822–1881), expressed the self-understanding of that social group. In the first issue of the new journal *Der Civilingenieur*, he described the "class of *Techniker*" along with the requirements of their education and training. The *Techniker*, he observed, stood between the traditional classes of scholars, artists, merchants, and entrepreneurs. Unlike the scholar, he possessed know-how and skills enabling him to construct machines or to produce new kinds of materials; but he also differed from the traditional practical man through his school-based *Bildung* (knowledge and education). Weber defined *Techniker* as "professionals" who possessed knowledge about mathematics as well as the laws of statics, mechanics, and chemical affinities, and who "applied" this knowledge in practical work in order to satisfy public and private "material needs." Application of scientific knowledge presupposed, however, its interconnection with technical knowledge and techniques. Therefore, Weber explained, scientific knowledge was merely "a basis" that had to be supplemented by "specialized training" (*spezielle Fachausbildung*) in view of the practical demands of the engineering profession.[44]

Hence, when Weber highlighted *Bildung* as the trademark of the engineer's profession, he changed and extended the older humanistic meaning of that term to include technological and scientific education as well as specialized training (*spezielle Fachausbildung*) or what was also designated as "*praktische Bildung*." As the historian Jürgen Kocka has pointed out, the engineers' emphasis on *Bildung* has to be put into the context of a society in which the humanistically educated elite, the *Bildungsbürgertum*, which despised commerce and industry, defined social prestige. In the eyes of the *Bildungsbürgertum*, the engineers and *Techniker* had a low social prestige because of their low *Bildung* in the sense of lack of humanistic general knowledge (*Allgemeinbildung*).[45]

THE FOUNDATION OF THE TECHNICAL UNIVERSITY OF BERLIN

It is beyond the scope of this book to detail the Industrial Institute's changes between the late 1820s and the 1850s, accompanying the shift toward the education and training of engineers and *Techniker*. The following overview may suffice to show that from a bird's-eye perspective there was not only fluctuation and change, but also an astonishing continuity both on the institutional level as well as with respect to the Industrial Institute's combination of scientific and technological instruction and of theory and practice. Moreover, the character of the institutional changes resembled the kind of changes that occurred at the Bauakademie, as discussed in chapter 15. In both institutions there was a raising of the level of scientific education and of technological specialization, even though the mathematical-scientific teaching at the Bauakademie was on a higher level, and only the Bauakademie presupposed an education at the *Gymnasium*.

In 1826, the upper class of the Industrial Institute was extended from a one-year course to a course of one and a half years; five years later, it was further extended to a two-year course. In parallel to this, the students' entrance age changed from fourteen to seventeen years. Three decades later, the average entrance age was nineteen, and many of the students then had first completed an apprenticeship. In the 1850s and 1860s, the majority of students had additionally attended a lower industrial school in the Prussian provinces (*Provinzial-Gewerbeschulen*) before they entered the Industrial Institute in Berlin. These changes reflect the alterations of the figure educated and trained at the Industrial Institute: it now was primarily the *Techniker* or "engineer" employed in a large, mechanized factory rather than the owner of a small workshop.[46]

In 1850, the state administration responded to these changes by a reorganization of the Industrial Institute. The lower class was disbanded, while the upper class was extended to a three-years course. Furthermore, students now had permission to attend lectures at the University of Berlin.[47] The stronger emphasis on more advanced scientific education went hand in hand with the specialization of technological education and training. While in the first part of the course all students still received the same basic instruction, in the second part they now had to choose between mechanics, chemistry, or the building craft.

In mechanics, students received specialized instruction on machines, ranging from materials for constructing machines and machine design all the way to the principles of steam engines and railways. Three days per week were reserved for practical work in the workshops. The teaching of chemistry now comprised lectures on analytical chemistry and chemical technology as well as extensive training in the laboratory, first two days per week and later daily. Future builders were instructed in technical drawing, architectural design, modeling, stone cutting, and some mechanical engineering. In October 1860, the specialized courses on chemistry were extended to include the science of metallurgy (*Hüttenkunde*), while the courses on building narrowed to shipbuilding. Future machine engineers now had the possibility of adding a whole year of practical training in the workshop after they had completed their three-year courses.[48]

In April 1866, the Industrial Institute was renamed the Industrial Academy (*Gewerbe-Akademie*). Around this time, debates about the unification of the Industrial Academy and the Bauakademie were in full swing. In 1846, former students of the Industrial Institute had created a new association, named Hütte, which in 1856 had been transformed into the Association of German Engineers (Verein deutscher Ingenieure). Its first director, Franz Grashof (1826–1893), since 1863 professor at the Polytechnikum in Karlsruhe, was a vigorous proponent of advanced polytechnic schools (*polytechnische Hochschulen*).[49] At a meeting of the Association in 1864, he argued for the unification of the Bauakademie and the Industrial Academy. A unified *"Technische Hochschule"* located in Berlin, he pointed out, would educate future engineers and *Techniker* for both civil service and private industry.[50] Fifteen years later, in 1879, his proposal became reality: the Industrial Academy was unified with the Bauakademie to become the Royal Technical University of Berlin (Königlich Technische Hochschule zu Berlin). In October 1916, the Mining Academy of Berlin, which had been founded in September 1860, was also integrated into the Technical University.

"ENGINEERING SCIENCE" AND "TECHNOLOGICAL SCIENCE"

These institutional changes went hand in hand with a discourse on the type of knowledge promoted at the Industrial Institute and the Bauakademie. In this discourse, the term "useful science" was slowly replaced by terms such as "scientific-technological" knowledge, "engineering science," and "technological science." For example, in 1846 the founders of the Hütte proclaimed that the association's goal was the promotion of "scientific-technological education and training" (*wissenschaftlich-technische Ausbildung*).[51] In 1870, the Industrial Academy introduced courses in "engineering sciences" (*Ingenieur-Wissenschaften*).[52] And when in 1864 Franz Grashof proposed the foundation of a technical university, he pointed out that it would comprise the "technological sciences" (*technische Wissenschaften*) and the "auxiliary sciences" (*Hilfswissenschaften*), that is, "mathematics, mechanics, and natural sciences [*Naturwissenschaften*]." He explained that the core subjects covered by the technological sciences were architecture, geodesy, hydraulic engineering, road construction, shipbuilding, mining and metallurgy, and chemical technology; to these one could add, depending on local circumstances, agriculture, forestry, and trade and commerce.[53] Simultaneously, he also used the terms "technological-scientific knowledge" (*technisch-wissenschaftliche Kenntnisse*) and "technological-scientific education and training" (*technisch-wissenschaftliche Ausbildung*).[54]

An important problem was the relation of theory and practice at technical universities. Grashof took it for granted that practical training was an important part of their teaching. A student entering a technical university, he pointed out, should first have completed one or two years of practical training in the profession of his choice. At the university, he would receive additional training in the drawing and design classroom, the laboratory, and the workshop. Grashof took particular pains to justify the necessity of workshops. "The usefulness of practical workshops," he stated, was obvious, in particular with respect to the students studying machine building. There was just one argument against workshops: the limited budget of a university. To this argument, Grashof's response was that model workshops could be established everywhere, as they were not expensive. With respect to other workshops, he added, there was "a general agreement" that a mechanical workshop of modest size was a necessity.[55] Thus, Grashof argued not only for the combination of "technological and auxiliary sciences," but also for the continuation of the combination of theory and practice, which had long been established at the mining academies, the Bauakademie, and the Industrial Institute of Berlin.

17 THE BIG PICTURE: USEFUL SCIENCE, TECHNOLOGICAL SCIENCE, TECHNOSCIENCE

In the previous chapters, we followed various activities of the practitioners of useful science, and in chapters 11 and 12 our analysis of programmatic texts on the sciences of mining and of saltworks yielded insights into the meaning of the concept of "practical" or "useful science." Furthermore, our discussion in chapters 15 and 16 of the Academy of Civil Engineering and Architecture and the Industrial Institute illuminated the historical shift in terminology from useful science toward nineteenth-century technological science. In this chapter we first summarize the main epistemological features of useful science, then discuss how they relate to nineteenth-century technological science and more recent technoscience. This discussion will bring us back to the overarching questions raised in the introduction concerning the deeper history of present technoscience and the big picture of the history of science.

The Prussian ministers, officials, and men of science who supported the useful sciences pursued the goal of promoting expertise for the management of natural resources and the improvement of industry. They intended, in particular, to educate and train expert officials for the technical departments of the state bureaucracy. Hence, the core disciplines of useful science related to demands of the Departments of Trade and Industry, Mining and Smelting Works, Building, and Forestry in the General Directory in Berlin (later renamed "sections" of the State Ministry) and the subordinate authorities in the provinces. In addition, the state also supported military schools, private schools for agriculture and forestry, and the Industrial Institute, which trained engineers and technical chemists for private industry. As we have seen from the examples of mining science and saltworks science, useful science systematically interconnected technological and scientific knowledge and inquiry. Hence, it differed significantly from the later concept of applied science, if "applied science" means natural science directly applied to industry. Instead, there have been more indirect routes for scientific knowledge to become useful. One of these was its coupling with technological knowledge in the institutional framework of useful science.

TEACHING AND RESEARCH IN USEFUL SCIENCE

Practitioners of useful science were engaged in both teaching and research. The teaching of future engineers and other technical experts certainly made up the core of their activities, whereas their research was, compared to today's research, a less frequent and more modest endeavor. Even so, practitioners of mining science set up laboratories, collections of naturalia, of drawings, and of three-dimensional models as well as model workshops for systematic scientific and technological experimentation. As Humboldt pointed out in his programmatic text on saltworks science (chapter 12), research in the context of the useful sciences was time-consuming and exploratory. This was a feature that the useful sciences shared with the experimental philosophy of the time, as well as with the modern natural sciences, although the overarching research aims of these types of science were not the same. In the framework of useful science, the general objective of research was to come to practically useful results, but this by no means excluded studies of the natural conditions of technical devices. In experimental philosophy, the most important research aim was the understanding of natural things and processes, with usefulness being a side effect. Aims certainly motivate research, but they do not determine it. If we consider research as a process that is also conditioned by conceptual and material resources, and by the researchers' ever-evolving concrete questions, overarching research aims are perhaps less important than is commonly believed. Aims typically vary in the course of the research process, along with improvement of under-standing. Hence, in the context of useful science explorative research could be as open-ended as in experimental philosophy or modern natural sci-ence. A case in point is Gerhard's series of experiments on useful materials, discussed in chapter 8, which mirrored academic chemists' experiments on natural substances.

With respect to teaching, useful science aimed at a combination of theory and practice. Theory taught in the classroom was supplemented by practice, that is, work in mines, smelting works, factories, farmlands, or at construction sites. In the ideal case, theory and practice were learned in par-allel. Hence, useful science also included elements of knowledge and things originating directly in local practice. In this respect, all of the Prussian insti-tutions of useful science followed the model of the mining academies. The Academy of Civil Engineering and Architecture and the Industrial Institute in Berlin—but also the schools of agriculture and forestry, the chemical-pharmaceutical boarding schools, the military schools, and to some extent even the young University of Berlin—emulated the mining academies' model for combining theory and practice. And all attempts to deviate from this model, such as Gerhard's plan to establish a mining academy in Berlin, were highly contested and destined to fail.[1] Furthermore, in the classroom, teaching the useful sciences went far beyond the traditional form of lectures. It included experimentation, technical drawing, mathematical tutorials as well as the use of maps, three-dimensional models, and collections of naturalia and of machines.

COMPONENTS OF KNOWLEDGE IN USEFUL SCIENCE

We now turn to the different components of knowledge in useful science. Practitioners of the useful sciences collected and generalized local technical knowledge embedded in practice, thereby creating the body of technological knowledge taught in the classroom and published in textbooks. They also performed experiments to acquire knowledge about imported steam engines, mining machinery, the preparation of new varieties of steel, ways to intensify the growth of crops, techniques to control insect pests, and so on. Thus, their objects of inquiry were processes of making and doing, the raw materials and instruments involved in these processes, and the finished products and artifacts; and the knowledge acquired in this experimental way was technological knowledge as well. Furthermore, practitioners of useful science incorporated parts of the natural sciences into their science. They did this with care, guided by the question of which specific parts of the natural sciences provided insight into the natural conditions of the technical things and processes in question. And if the required natural-scientific knowledge was not yet available, they undertook their own natural inquiry. For example, they studied the geognostic structure and landscape in a mining region, thus contributing to the development of geognosy and physical geography; or they studied the behavior of an imported plant on experimental farmland or the reproduction rate of insects in a cultivated robinia forest, thus contributing to botany or zoology. All useful sciences of the late eighteenth century and early nineteenth century included significant components of natural science, and all of them spurred natural inquiry.

How did the technological and natural-scientific knowledge components of the useful sciences relate to each other? In some cases, studies of technological and natural-scientific knowledge merely interacted while preserving their own integrity. For example, there was cross-fertilization between geognosy, mine surveying, and the construction of mine plans, but these components of mining science also had some degree of independence from each other. In other cases there was strong interdependence, or even convergence, of natural-scientific and technological knowledge and inquiry. An example of the latter is the introduction of the water column machine, which went hand in hand with experimental, mathematical, and physical-theoretical studies. In his textbook on mining machines from 1763 Henning Calvör thus designates his chapter on the water column machine as a "theoretical and practical description."[2] Another example of convergence is Gerhard's exploration of preparations of new kinds of alloys, ceramics, and glass. In these experiments, Gerhard extended scientific chemical knowledge about the reactivity of metals and earths to new substances, and at the same time explored the possibility of producing new and potentially useful materials. Humboldt's invention of a miner's lamp and respiration machine, detailed in chapter 10, is yet another example of interaction and local convergence of natural-scientific and technological inquiry. In his work of invention, Humboldt often tinkered with technical constructions without asking scientific questions. But he also incorporated oxygen, eudiometers, and chemical techniques, developed in the institutional context of academic pneumatic chemistry, into this work. What is more, parallel to his work of invention he performed chemical analyses of airs (gases) in mines. This is a prime example of interaction between scientific and technological inquiry. At the end of his invention, technological and natural-scientific inquiry even converged, as when Humboldt used the eudiometer to monitor the functioning of his new miners' lamp and simultaneously for analyzing the air in a mine.

A COHERENT SYSTEM OF KNOWLEDGE?

The useful sciences consisted of different components of knowledge. This raises the further question of how these components were related to each other in a particular useful science. Did they share a common epistemic foundation? Did mining science, for example, have basic principles that rendered it a coherent system of knowledge? The answer to this question is no. The selection and assemblage of the components of knowledge that constituted mining science resulted from the particular challenges of mining and the actors' objective of acquiring knowledge for the improvement of mining industry. Just as the single components of mining science were tailored to the particular technical field they referred to, their interconnection was also guided by considerations of practical relevance. It was the pragmatic perspective of the mining industry that prompted the practitioners of mining science to assemble such diverse components as knowledge about surveying, assaying, water wheel systems, construction of shafts, as well as geometry, mineralogy, geognosy, metallurgical chemistry, and hydraulics. While there was epistemic interdependence, and even local convergence, among particular components of mining science, there was neither an overriding logic nor a deductive structure nor any set of basic principles that might tie everything together completely and thus generate a scientific system comparable, for example, to Newtonian mechanics. On the contrary, from an epistemological point of view there were gaps between certain knowledge components of mining science. For example, there was no cross-fertilization between geognosy and mining engineering, or between chemical metallurgy and prospecting. Like other useful sciences, mining science was a true "system" of knowledge only from a pragmatic point of view, namely insofar as it assembled knowledge components relevant to a distinct field of practice.

In line with this, practitioners of useful science did not attempt to press all components of knowledge into the corset of a uniform language, but rather used different languages and formulas to represent different subfields of a useful science. Nor did they attempt to elaborate a general, unified theory of mining, or of metallurgical chemistry, agricultural science, forest science, and so on, let alone a unified theory of technology comparable to the eighteenth-century unified philosophies of nature. Attempts to formulate unified theories of nature presuppose the belief that nature is structured throughout in a regular, lawlike, or even reasonable way. An analogous belief did not exist, and does not exist, with respect to technology.

PRAGMATISM AND A NEW CONCEPT OF SCIENCE

Empirical and methodical knowledge certainly were the most significant types of knowledge within the framework of useful science, and they stemmed from studies of technical things and processes as well as from the natural sciences. Needless to say, empirical knowledge goes beyond sensual perception and also includes results of cognitive processes, which create links between phenomena abstracted from local peculiarities, and thus generate general concepts. We have already mentioned that the early modern technological books presented accumulated empirical knowledge about technical practices. For the practitioners of useful science, these books were important reservoirs of knowledge, as were publications on chemistry, mineralogy, botany, geometry, hydromechanics, and so on. The natural sciences offered, in particular, advanced empirical knowledge and methods such as note taking in experimentation and quantitative chemical analysis. As we have seen above, scientific theory, such as hydromechanics, the chemical theory of oxidation, and theories of the history of the Earth, was integrated into useful science as well. Yet the type of theory at stake here is worth noting: These were special theories that each referred to a well-defined, limited scope of objects of inquiry.

If there was something like an epistemology of useful science, it was pragmatism. The protagonists of useful science considered their science not as an end in itself but as related to practice. We have detailed this issue in chapter 11, in connection with an examination of Heinitz's epistemic values. Heinitz emphasized not "truths" but rather "reliability of knowledge." "Reliable knowledge" implied that knowledge was given a certain function in practice. It enabled men to get things right. This attitude differed profoundly from early modern natural philosophers' search for universal truths, articulated in high philosophies of nature. It also accorded with the fact that practitioners of useful science did not attempt to create deductive systems of knowledge.

All in all, useful science represented a new concept of science, which partly overlapped with contemporary natural philosophy but also differed from it in important respects. Both types of science produced and transmitted knowledge, that is, they organized research as well as teaching, and they elaborated methods and values for these purposes. But whereas natural philosophy focused on knowledge about nature, useful science aimed at combining natural and technological knowledge and inquiry. Further, natural philosophy highlighted knowledge acquired for its own sake, certainty of knowledge, truth to nature, and high theory. By contrast useful science ultimately aimed at making knowledge useful for practice, endorsed a pragmatic epistemology, and eschewed high theory.

NINETEENTH-CENTURY TECHNOLOGICAL SCIENCE

Some fifty years ago, the American historian of technology Edwin Layton argued in a widely lauded essay that the modern engineering and technological sciences, especially the engineering sciences in the US, were the result of a "scientific revolution" in late nineteenth-century technology, and that in previous centuries technical knowledge had still been firmly entrenched in the arts and crafts.[3] Historians of technology studying the developing technological knowledge in Europe, however, regard the nineteenth-century technological sciences not as the outcome of a sudden revolution, but rather as the result of a continuous evolution.[4] In line with the latter view, the previous chapters have provided evidence that as early as the second half of the eighteenth century the useful sciences, institutionalized in the German states, established a scientific approach to technological problems. Like eighteenth-century French engineering science, the useful sciences embodied a new concept of science that comprised technological as well as natural-scientific knowledge. In the following, we compare eighteenth-century useful science with nineteenth-century technological science. We will make this comparison from an epistemological point of view, focusing on modes of inquiry, methods, and constellations of natural-scientific and technological knowledge. Thus we will abstract here from the peculiar local features of technological sciences in different countries and at different institutions. Debates among historians of technology concerning the characterization of engineering and technological science serve as a reference point for our comparison.

In the 1970s, Layton characterized the natural and the technological sciences as "mirror-image twins," drawing attention to their shared features but also to their autonomy.[5] This characterization was not least an objection to the long-dominant linear model, according to which technology only applies the results generated by the natural sciences without making significant changes or contributions to them. By contrast, Layton made an appeal for the equal status of technological and natural sciences. Based on this, he proposed to replace the linear model with an interactive model of science and technology. Concerning the shared features of the natural and the technological sciences, he pointed first of all to the similarity of their methods. The development of the technological sciences in the second half of the nineteenth century, he asserted, "involved the adoption by engineers of the theoretical and experimental methods of science." He highlighted especially the use of mathematics and the establishment of programs of "basic research in laboratories."[6]

In the case of the nineteenth-century engineering science concerned with the strength of materials (by far his most elaborate example), he observed that physicists and engineers at first shared their methods as well as their theories and objects of inquiry—that is, materials and their properties such as strength and elasticity. But between the 1830s and the 1870s, the differences between them increased, with the result that by the 1880s there were two different communities of engineers and physicists. Simultaneously, the engineering science concerned with the strength of materials assumed the characteristics of an autonomous "sister discipline" to that of physics.[7] What were the differences between these sister disciplines? In Layton's eyes, the main difference consisted of the role played by fundamental theories.

While physicists "tended to explain their findings by reference to the most fundamental entities, such as atoms, ether, and forces," academic engineers were content with low theory and models involving macroscopic entities—for example, modeling a beam as a bundle of fibers.[8] This was related to considerations about the usefulness of materials, which took up more room in the context of the evolving engineering science than in the context of physics. Another difference, Layton asserted, concerned the implementation of mathematics. Physicists normally admitted only rigorous mathematical methods, but academic engineers tended to rely more strongly on approximations and less rigorous graphical methods.[9]

Hence, with respect to physicists' and engineers' studies of the strength of materials, Layton came to conclusions that largely coincide with our analysis of useful science. First, he pointed out that physicists' and engineers' objects of inquiry overlapped; in other words, natural-scientific and technological inquiry interdepended or even converged. Second, both the engineering sciences and the useful sciences organized research, and they both established research laboratories and adopted research methods from the natural sciences. Third, usefulness of knowledge was the ultimate aim in the engineering sciences and the useful sciences. Fourth, the practitioners both of engineering science and of useful science eschewed the type of high theory elaborated in physics and in natural philosophy, respectively.

Layton's view of the natural and technological sciences as "mirror-image twins" highlighted the fact that the two types of science share certain values—especially the mathematical approach and the experimental method—but prioritize them in reverse order. In accordance with what we have shown for the useful sciences, the concept of mirror-image twins allows all kinds of interactions between the twins, ranging from weak interaction to local convergence of natural-scientific and technological inquiry. Hence, Layton pointed out that in the late nineteenth century, when the natural sciences and the technological sciences had been separated institutionally, there were still persons who were members of both the community of scientists and that of academic engineers. He designated these intermediaries as "engineer-scientists."[10] Further, he concluded that technological science was "an intermediary body of knowledge which connects science and technology." David Channell called this view the "interactive model" of the relationship of science and technology, which Layton proposed as a replacement for the traditional linear model.[11]

In a pioneering study of the development of electrical engineering (*Elektrotechnik*) as a scientific discipline in Germany in the last two decades of the nineteenth century, the German historian of technology Wolfgang König arrived at similar conclusions to Layton's. König showed that academic electrical engineering was a science of both electricity and electrical machines. It thus emerged at the intersection of the natural science of physics and mechanical engineering. Academic "electrical engineering," König stated, came into being in the "field of interplay of scientific and technological disciplines." Accordingly, German technical universities offered courses of study that included mathematics, the natural sciences, knowledge about machine construction, and further technological issues.[12] As in the development of studies of the strength of materials, the young scientific discipline of electrical engineering was soon divided into more specialized subdisciplines. Around 1900, there were different chairs for "theoretical electrical engineering" (*theoretische Elektrotechnik*), construction of electrical machines, and electrical supply systems. Theory-oriented electrical engineering put a relatively strong emphasis on mathematics, physics, and chemistry, while practice-oriented electrical engineering emphasized technical issues and demands of industry. The proportion of the two branches differed among German technical universities, but all universities offered some parts of them. Further, König pointed out that by the end of the nineteenth century, technical universities increasingly hired professors of electrical engineering who had acquired practical experience in industry. Hence, like the useful sciences in the decades around 1800, academic electrical engineering remained open to the influx of practical-technical knowledge originating in industry.

Summing up, König identified the following five characteristics of modern technological sciences:[13]

- Their central aim is the description and explanation of existing and exploration of possible technical practices.

- They organize research as well as the professional education and training of engineers and *Techniker*, who in the German states up to the 1870s were mainly public servants.

- They develop a multiplicity of methods, especially the use of models. Models reduce the complexity of the structure and function of technical objects, and thereby allow the formulation of valid, abstract rules concerning technical objects.

- In the technological sciences, distinct cognitive elements are interconnected: accumulated experience from industrial practice; knowledge and rules acquired through studies of technical objects, including the use of models; natural laws; natural-scientific theories; and mathematics.

- There is no common theoretical and methodological foundation of any technological science, nor is there one comprehensive technological science.

Hence, both Layton's and König's characterization of the late nineteenth-century technological sciences coincide with the characterization of the useful sciences presented above. From an epistemological perspective, the useful sciences and the modern technological sciences share many features. Clearly, there are also differences between them. The useful sciences were oriented toward the state, while the technological sciences became increasingly relevant for private industry. Furthermore, as a result of the development of the modern system of specialized scientific disciplines, natural inquiry was moved to the theoretical subdisciplines of technological sciences, while in the framework of the useful sciences this specialization did not yet exist. Eighteenth-century mining science, to name one example, thus contained more natural-scientific parts than late nineteenth-century mining science. In the second half of the nineteenth century, the technological sciences also received their modern, codified guise—obligatory courses of study, curricula, examination regulations, canonical textbooks and handbooks, specialized journals, professional meetings, interest groups, and so on.[14] Thus, for the historian numerous additional questions could be asked concerning the continuity and discontinuity of the long-term historical process in question. Even so, there is strong evidence that in Germany the useful sciences developed gradually into the modern technological sciences and their neighboring natural-scientific disciplines.

TWENTIETH-CENTURY TECHNOSCIENCE

We now turn to recent technosciences, tackling the question of whether they present a novel constellation of knowledge and a novel mode of doing science, compared to the eighteenth-century useful sciences and the nineteenth-century technological sciences. As we saw in the introduction, many historians assume that technoscience is a novelty of the twentieth century and that pure, disinterested science prevailed prior to this. As an example of technoscience we take biotechnology, focusing on in vitro cell cultures.

In her pathbreaking study *Culturing Life*, the historian and anthropologist Hannah Landecker has traced the ways in which artificial forms of cellular life were created over the course of the twentieth century.[15] Landecker starts her story by showing how the American embryologist Ross Harrison succeeded for the first time in cultivating tissue fragments for several weeks in a suitable medium. In 1907, Harrison grew a nerve out of a fragment of embryonic frog tissue. Because he wanted to study embryonic nerve development, in vitro cell cultures were first a new technique in the context of natural science. Harrison's goal was the observation of nerve development, and the novelty of his cell culture was that this could now be done outside the body. Even though the new technique was a crucial component of his research, it could be clearly identified as a technique or "technical object"—as distinguished from the main "epistemic object," which was the natural development of nerves.[16] However, other life scientists immediately changed this relationship.

The Franco-American surgeon Alexis Carrel, in particular, became interested in the technique itself and the range of its possibilities. He wanted to know whether it was possible to stimulate cell division in vitro over an extended time period. Working with embryonic chicken heart cells, he developed a new apparatus of tissue culture after 1918, including glassware and microcinematography, and manipulated the nutritive medium in which he kept cells alive and growing. Much of his effort went into the exploration of the new apparatus. Landecker introduces a challenging epistemological twist when she calls his technological inquiry "open-ended" and contrasts it with the "highly controlled, hypothesis-driven" experiments of a natural scientist like Harrison. "Carrel was a surgeon," she observes, "with a much stronger tendency to tinker with tissues in an open-ended way—to see how far one could push them and what would happen when one did—than to experiment in a highly controlled, hypothesis-driven way."[17] But Carrel was by no means a mere tinkerer. He was also interested in the question of "permanent life" or potential immortality of cells in culture. Thus, in his work technological and natural-scientific inquiry partially overlapped. This was a major step toward the formation of biotechnology.[18]

Goals of application of cell cultures, Landecker points out, were first articulated in the 1940s, when human cells could be cultivated on a large scale. Human tissue culture was then used to grow viruses in order to find a vaccine for polio. The disembodied human cell culture as "productive technology came into being."[19] Again, the development of this new technology implied that natural-scientific knowledge about the growth of viruses and knowledge about the techniques of maintaining cell cultures were interdependent. In the 1950s, techniques of freezing cell cultures opened the way to artificial insemination in agriculture. A decade later, technical interventions into cell cultures had reached a new level. Landecker argues that biology then had become a synthesizing science, comparable to synthetic chemistry. Chemistry had long been a science that merged scientific inquiry into the nature of chemical reactions with inquiry into the technical conditions of creating new kinds of substances. In nineteenth-century organic chemical synthesis, the distinction between discovery and invention no longer made any sense.[20] In a similar way, biologists now began to synthesize cells. Fragments of cells were used to reconstitute whole cells, and somatic cells from different species of animals were fused together into new kinds of hybrid cells. There can be little doubt that these experiments undermined not only the distinction between living things and technical artifacts, but also that between biological and technological research. Studies of cell cultures had developed into a biotechnology and, in more general terms, into "technoscience."

MEANINGS OF TECHNOSCIENCE

The term "technoscience" became popular in the late 1980s through Bruno Latour's *Science in Action*. Latour used "technoscience" as a shorthand for the interaction of "technology and science" in twentieth-century research institutions, in which experimental research typically depends on a whole range of machines and computers as well as on cooperation between scientists with engineers. He also used it to characterize the innovative, problem-solving work done by academically trained staff in high-tech companies.[21] As Latour was silent about the boundaries of technoscience, critics from all corners soon objected that technoscience was not a well-defined term, and that not all modern science is technoscience. Physical research with the Large Hadron Collider at CERN is a prime example of research relying on big apparatus and teams of scientists and engineers, but the vast majority of scientists and scholars would certainly not regard it as a good example of technoscience. On the contrary, they would argue that research on Higgs bosons is a typical case of basic research.[22] Generally speaking, all experimental sciences depend on instruments, apparatus, materials, techniques, and technically trained experimenters. Historians of science hardly question this kind of interdependence of science and technology, but most of them rightly reject the conclusion that all experimental sciences are technosciences.

It should be noted, however, that the ambiguities of the term "technoscience" mirror the ambiguities of the older and widely used term "technology." In English, "technology" means, first, practices of doing and making, and the ensemble of processes and artifacts that humans have brought into existence; second, the ways in which things and processes are commonly made and done, that is, techniques and know-how; and, third, systematic technological knowledge about these subjects, which includes the institutionalized engineering and technological sciences. As Edwin Layton pointed out almost half a century ago, the third definition of technology as "systematic knowledge of the industrial arts" is often lumped together with technology in the sense of "techniques" and "the things produced by techniques." Likewise, the academic engineer or "technologist" is often confused with the practical engineer or "technician."[23] This has consequences for the definition of "technoscience."

Depending on what is meant by "technology," the meaning of "technoscience" varies considerably. It may mean, as has just been pointed out, that experiments depend on technical equipment, and, inversely, that experimental technical equipment evolves in the context of experimental research; and this weak definition of technoscience may further be extended to the experimenters' skill and technical knowledge. By contrast, based on the meaning of technology as systematic technological knowledge and technological science, it may mean interdependence and local convergence of technological and natural-scientific knowledge and inquiry in the framework of distinct institutions, such as military schools, engineering colleges, technical universities, modern research universities, and industrial research labs. This second, stronger meaning of technoscience is the one adopted in this book, and the biotechnological research described above is an example of technoscience in this stronger sense. Our example of biotechnology also shows that application is within the horizon of possibilities of technoscience but not always its immediate aim.

In the vein of Latour's extension of "technoscience" to innovative work in high-tech companies, "technoscience" is also defined as the entanglement of science, technology, and industry. This third meaning of "technoscience"—which is not adopted in this book—hinges on the meaning of "technology" as industrial practices of doing and making. "Technoscience" then extends to industrial production as well, the implication being that high-tech industry embodies natural-scientific and technological knowledge and that institutional boundaries between research organizations and industry are irrelevant. In his recent book *A History of Technoscience*, David Channell highlights an example of technoscience in this third sense, when he points out that the production of atomic weapons in the United States after World War II was "one of the first areas in which the military-industrial-academic complex merged with big science."[24] Channell has a point here, but the question arises of whether this example can be generalized. Channell's own work also demonstrates that there are numerous examples of transition from research institutions to industrial production sites that are significantly less smooth and instead involve new challenges. The modern natural and technological sciences are firmly embedded in very specific research and teaching institutions that differ significantly from the institutional framework of industrial production; the relationships between these two types of institution should be an issue of targeted historical research.

The philosopher and media theorist Jutta Weber goes a step further than Channell, extending the reference of technoscience to include the market and everyday consumption. Weber argues that since the 1980s there has been an "increasing hybridization of science, technology, industry and society." Considering the case of robotics and robots, she states that it teaches us to understand "technoscience as cultural and everyday practice." Here the erasure of boundaries between different social institutions and practices is almost complete. What should be seen as relationships—namely, relationships between institutionalized research on intelligent robots, commercial production of robots, the consumer market, and everyday uses of intelligent robots in the kitchen and dining room—is construed as a single whole or one "ensemble of cultural practices."[25] Needless to say, this view of technoscience bars any attempt to understand the impact of social institutions on knowing and doing as well as problems of transmission across institutional boundaries.

CONCLUSION

In the introduction, we raised the question of whether recent technoscience has a deeper history. Our analysis leads unambiguously to the conclusion that this is the case, and that the argument that technoscience is a radical novelty that breaks with the values of past science is no longer tenable.[26] Based on our definition of technoscience as sustained coupling of natural-scientific and technological knowledge and inquiry in the context of specific institutions, the eighteenth-century useful sciences and the nineteenth-century technological sciences were early modes of technoscience which are fully comparable to biotechnology, nanotechnology, genetic engineering, robotics, and other more recent technosciences. If we also take into account the numerous projects carried out by scientific-technological experts in the centuries before, mostly outside of academic institutions, the roots of technoscience go even deeper in history. Modern scientists and their early modern predecessors have not only asked what exists in nature and what we can discover, but also what it is possible to do and to make. Conversely, engineers have not only asked what artifacts can be invented, but also what the natural conditions of inventions are. Disinterested studies of nature in the framework of ancient philosophical schools and university-based natural philosophy certainly played an important role in the long process that eventually brought our modern system of scientific disciplines into existence. But natural philosophy was by no means the most important predecessor of the modern sciences. There were many additional knowledge traditions such as practical mathematics, mechanics, and chemistry as well as the eighteenth-century useful sciences and the nineteenth-century technological sciences that interconnected natural-scientific and technological inquiry and contributed significantly to the evolving system of sciences.

Late twentieth-century and twenty-first century high-tech industry, along with its global expansion and predominance over all other forms of production, is indeed a historical novelty, which depends, however, not only on science or technoscience but also on numerous additional factors besides science, not least on global capitalism and geopolitical factors. Recent technoscience also depends on factors other than industry, even though the influx of private capital into scientific institutions and the patenting and commercialization of scientific research have reached an unprecedented level. To put it briefly, technoscience and industry are not the same institution. Instead of lumping different societal practices and institutions together, we should see technoscience and industry as coevolving entities.

Today, in the age of the Anthropocene, not only industry but also large parts of everyday life—and even the life of our planet—depend crucially on the natural and technological sciences.[27] There are now only a few scientists who would claim that the goal of science is acquisition of knowledge for its own sake. The majority of scientists rather believe that the ultimate goal of science is practical usefulness, including the preservation of our planet. As the previous stories have shown, practical usefulness does not always serve corrupt goals of money making. Likewise, the interconnection of scientific and technological inquiry is not necessarily a disreputable endeavor that would imply a radical break with the ideals and values of science. On the contrary, before the Industrial Revolution and capitalism reached their first peak in the late nineteenth century, scientists often pursued humanitarian goals alongside the goal of practical usefulness. It is time to rethink our understanding of science to include not only natural but also technological knowledge as well as traditions of technoscientific inquiry, which for ideological reasons were long actively excluded from the big picture of the history of science.

NOTES

Introduction

1. For a discussion of the term "applied science" see Bud 2012; Bud 2014; Gooday 2012; Lucier 2012. For "fundamental research" and "basic research" see Clarke 2010; Schauz and Kaldeway 2018.

2. For the concept of mode-two science, see Nowotny, Scott, and Gibbons 2001; for the epochal-break thesis, see Nordmann 2011; for the argument that there is an "epochal elevation" of technology in the framework of technoscience and that technoscience is the result of postmodernity, see Forman 2007, 1. For further literature on "technoscience," see chapter 17.

3. See Daston and Galison 2007; Gaukroger 2006; Gaukroger 2010; Gaukroger 2016.

4. See Dear 1995. By contrast, Schuster defines mixed mathematics as a "contested borderland between natural philosophy and practical mathematics" (Schuster 2017, 47).

5. By contrast, the French term *la science* and the German term *Wissenschaft* comprise all academic disciplines including the technological sciences, the social sciences, and the humanities. See Olesko 2020.

6. Park and Daston 2006, 2–3, 4; Blair 2006, 365.

7. Cunningham and Williams 1993, 421. See also the discussions in Kohler and Olesko 2012; Daston 2017.

8. On early modern practical chemistry see Klein 1994; Klein 2005a; Multhauf 1966. On practical mathematics see Cormack, Walton, and Schuster 2017; Schemmel 2008.

9. It should be noted that the term "arts and crafts" comprises agriculture and household production. For the concept of knowledge economy used here, see Renn 2020.

10. See Merton 1970 (1938); Long 2011; Renn 2020; Alder 1997.

11. Layton 1971.

12. See chapter 17.

13. For a recent attempt to elaborate an alternative big picture, see Renn 2020.

14. In the past twenty years, the literature concerned with an alternative to the prevailing grand narrative has grown enormously. Hence, a full overview would go beyond the scope of this essay. To these more recent publications belong the following monographs and essay collections: Ash 2010b; Belhoste 2011; Bertucci 2017; Cook 2007; Cormack, Walton, and Schuster 2017; Dupré and Lüthy 2011; Jacob 2014; Jacob and Stewart 2004; Jones 2008; Jones 2016; Klein 2012b; Klein 2015; Klein 2016a; Klein and Lefèvre 2007; Klein and Spary 2010a; Laboulais 2012; Langins 2004; Lefèvre 2004; Long 2011; Nummedal 2007; Popplow 2010a; Roberts, Schaffer, and Dear 2007; Renn 2020; Schemmel 2008; Roberts and Werrett 2017; Smith 2004; Valleriani 2010; Valleriani 2017; Tomic 2010; Tomory 2012; Wakefield 2009; Werrett 2010.

15. For an overview of attempts to manage natural resources in the German states, see Meyer and Popplow 2004.

16. For the history of this term and concept, see also Schatzberg 2018.

17. For the Royal Military Academy, established at Woolwich in 1741, see Mauskopf 2010. For the School of Naval Architecture set up at Portsmouth Naval Yard in 1811, see Schaffer 2007. For studies of French technological schools see Alder 1997; Belhoste 2003; Jones 2016; Langins 2004; Picon 1992. As the related studies on Prussia concentrate on single institutions, they will be quoted in their corresponding chapters below.

18. In the following, the term "useful science" will be used, since it occurs in the documents more often than "practical science."

19. See also Wise 2018. By contrast, in Phillips's recent study of the development of the natural sciences in Germany between 1770 and 1850, the focus is on universities, private learned societies, and sites of popularization of science, while all institutions promoting the alliance of natural and technological inquiry are excluded; see Phillips 2012.

20. See, in particular, Alder 1997; Langins 2004; Picon 1992; Schemmel 2008.

21. See Bleker, Lennig, and Schnalke 2018; Lehmann 1936.

22. Quoted in Schickert 1895, 18.

23. Rieck 1928. This building still exists at what is now the campus of the Charité.

24. See Poten 1896, vol. 4; Scharfenort 1910; Wise 2018.

25. For the figure of German romantic *Naturforscher* see Caneva 1997; Engelhardt 1990; Klengel 2010.

26. For the term "engineer-scientist," see Layton 1971, 578. See also chapters 13 and 17.

27. For an overview see Ashworth 2017.

28. Jacob 1997; Jacob 2014; Jacob and Stewart 2004; Mokyr 1990; Mokyr 2002; Mokyr 2005; Stewart 1992. In the 1950s and 1960s, Musson and Robinson came to similar conclusions, but their research was stigmatized as Marxist; see Musson and Robinson 1969. See also Inkster 1991; Inkster 2012.

29. Jones 2008; Miller 2009; Tomory 2012; Tomory 2014. See also Davids 2012; Inkster 2012; Jones 2016; Smith and Wise 1989.

30. Jones 2008, 17–18.

31. The period from 1835 until 1845 has been identified as the first phase of the establishment of the German machine building industry; see Lundgreen 1975, 200; Schröter 1962.

32. There were two centers of power in eighteenth-century Prussia, the king with his secret cabinet in Potsdam, on the one hand, and the General Directory (General-Ober-Finanz-Kriegs- und Domänen-Direktorium) in Berlin, on the other. The General Directory existed from 1723 until 1808, when it was replaced by the Staatsministerium. Although in the framework of absolutism the king made all final decisions, there was considerable room for the ministers' policy and tactics. For an overview of the Prussian state bureaucracy and governmental system in the second half of the eighteenth century and the early nineteenth century, see Brose 1993; Johnson 1975.

33. On the problem of shortage of wood see Radkau 1989; Radkau and Schäfer 1987.

34. Hahn 2011 (1998), 21.

35. On the Seehandlung, see Henderson 1958, 119–147; Radtke 1981.

36. For the takeoff thesis, see Rostow 1960. For the view that industrialization in Prussia was a slow, continuous process, in which different stages can be discerned, see Henning 1973; Radkau 1989. For an overview of different approaches to Prussian industrialization, see also Condrau 2005; Hahn 2011 (1998); Treue 1984.

37. See, for example, Brose 1993; Henderson 1958; Kocka 1969; Koselleck 1967. It should be noted that Brose studies the concept of "state-driven industrialization" and "images of the Prussian state as a great, modernizing economic leader" from the perspective of ideology (Brose 1993, 7, 5). He criticizes more or less all studies concerned with the role played by the state in Prussia's industrialization as being ideologically motivated. However, the Prussian state's technical activities and its promotion of technical expertise are not part of his story.

38. "Cameralism" was a seventeenth-century and eighteenth-century doctrine, mainly in the German states, which highlighted the role played by knowledge in practice. It defined the state authorities as the major promoter of useful knowledge

and thus of the improvement of practice and the common good. "Cameral science" was a university discipline introduced for the academic education of state officials. It compiled knowledge drawn from social, political, and economic theory, fiscal and legal knowledge, rules of administration, as well as some mathematics and natural science. See also the concluding parts of chapter 12.

39. The *Ingenieur* was a person concerned with mechanical engineering, whereas the term *Techniker* was broader, including, for example, technical chemists as well.

Chapter 1

1. Leibniz's memorandum is included in full in Harnack 1900, vol. 2, 76–78, here 77. The full text of the charter is in Harnack 1900, vol. 1.1, 93.

2. By contrast, in France the natural sciences and the philological and historical disciplines were represented in two separate academies, the Académie Royale des Sciences, founded in 1666, and the Académie Royale des Inscriptions et Médailles, founded in 1701. The latter was renamed the Académie Royale des Inscriptions et Belles-Lettres in 1716. The Royal Society, founded in 1660 in London, also dealt exclusively with scientific, mathematical, and technological subjects.

3. Euler worked from 1741 to 1766 at the Royal Prussian Academy of Sciences, then went to St. Petersburg. His often-cited attempt to install waterspouts in the fountains of Sanssouci, by having water pumped from the lower-lying Havel to a basin on a hill, is not the only example of his practical engagement. As more recent studies have shown, the failure of this venture was not Euler's fault but that of the practitioners; see Eckert 2002; Epple 2008.

4. See Eichler 1974; Henderson 1961; Olesko 2009; Olesko 2020; Strecke 2000; Strecke, Baier, Blauert, et al. 2000, 93–98.

5. See Olesko 2009.

6. See Strecke 2000, 86–91.

7. Archive of the Berlin-Brandenburg Academy of Sciences (hereafter cited as ABBAW) I-V-73 (Acta betr. die vom König befohlene Anstellung einiger Versuche zur Verbesserung des Kalks und Mörtels).

8. See Duffy 1974.

9. Eytelwein's activities will be discussed in chapter 15.

10. ABBAW I-V-80 (Acta wegen Untersuchung eines von dem Kaufmann Müncheberg allhier Fabricirten und gut befundenen Eßigs).

11. ABBAW I-V-72 (Acta betr. die Anlegung eines so genanten Conducteur oder Wetter-Ableiter zur Sicherheit des Vorraths-Magazin-Gebäudes am Schlesischen Thore), 6–18.

12. Lehmann was only an honorary mining councilor.

13. On Heinitz's professional career, see Weber 1976.

14. Radkau 1989, 60. See also Bayerl 1999; Meyer and Popplow 2004; Radkau and Schäfer 1987; Popplow 2010a.

15. Carlowitz 1713, 105.

16. Geheimes Staatsarchiv Preußischer Kulturbesitz (hereafter cited as GStA PK), I. HA, Rep. 121, Ministerium für Handel und Gewerbe. Abt. Bergwerks-, Hütten- und Salinenwesen no. 7958, 21.

17. Gleditsch 1774–1775.

18. GStA PK, I-XIV-43 (Acta die Botanischen Garten-Angelegenheiten betreffend 1801), 34–35.

19. In November 1803, Willdenow gave an account of trials in a concluding report, entitled "On the Cultivation of Potatoes" ("Über den Anbau von Kartoffeln"). ABBAW I-XIII-22, vol. III, 110; *Registres* of the Academy of Sciences 1801–1805; ABBAW I-IV-34, 42, 66.

20. See Klein 2015, 233–237. For a more comprehensive scientific biography of Achard, see Müller 2002.

21. On this discovery, see Klein 2015, 42–45.

22. See Müller 2002. Today, the village is part of Marzahn-Hellersdorf, a district of Berlin.

23. ABBAW I-IV-33, *Registres* of the Academy from August 24, 1786, until December 18, 1800, 195, 204, 253, 281. On the German economic societies see Popplow 2010b. The Märkische Ökonomische Gesellschaft zu Potsdam had been founded in 1791.

24. Achard 1796; Achard 1797.

25. Quoted in Müller 2002, 544.

26. ABBAW I-XIII-26 (Acta wegen dem Herrn Director Achard bewilligten Laboratorii zur Fabrication des Zuckers aus Runkelrüben), 1.

27. Quoted in Müller 2002, 291.

28. ABBAW, 1-XIII-26, 11.

29. Harnack 1900, vol. 1.1, 68–69.

30. For the following see ABBAW I-XIII-20, 9–16. See also Klein 2008a; Klein 2008b.

31. ABBAW I-XIII-20, 10–11.

32. Ibid., 7.

33. Ibid., 3–4.

34. ABBAW I-XVI-134, 29.

35. ABBAW I-XVI-122, 22; ABBAW I-XVI-131, 27.

36. ABBAW I-XVI-128, 26 (for 1762); ABBAW I-XVI-129, 26 (for 1763).

37. ABBAW I-XIII-26, 131.

38. See chapter 14.

39. Quoted in Harnack 1900, vol. 2, 338.

Chapter 2

1. Fischer 1820, 15–16.

2. For further details of Klaproth's biography, see Dann 1958; Klein 2015.

3. Anonymous 1784.

4. See chapter 14.

5. See chapter 4.

6. Klaproth gave an account of this journey in Klaproth 1789c. It is documented that he obtained the sample of Ceylonese zircon from Werner, the analysis of which led to the discovery of *"Zirkonerde"* (zirconium earth, today zirconia), announced in the same year as his discovery of uranium; see Hoppe 1989.

7. Kopp 1843–1847, vol. 4, 82.

8. Klaproth 1789a; Klaproth 1789b.

9. On anayltical chemistry in the late eighteenth century, see also Homburg 1999; Szabadváry 1966.

10. Klaproth 1789a, 390.

11. Klaproth 1789a, 391. Affinity tables presented the regularities in precipitations and other kinds of displacement reactions. On this issue, see Duncan 1996; Klein 1994; Klein 1995; Klein and Lefèvre 2007.

12. Kopp 1843–1847, vol. 4, 78.

13. Klaproth 1789b, 165.

14. According to the phlogiston theory, all base metals consisted of a variable metal calx and an invisible phlogiston, which was released in combustion and calcination (later termed oxidization). Bergman had proposed further that metals also release

their phlogiston when dissolved in mineral acids, so that a metal calx also resulted from acidic dissolution.

15. Klaproth 1789a, 394.

16. In the context of the phlogiston theory, the metal calx was a simpler substance than a metal, which was a compound. According to the rules of naming, names of the compound metal should therefore be derived from the metal calx. However, in the tradition of mining and mineralogy, metals were considered the more important substances. See Klein 2014a.

17. Kopp 1843–1847, vol. 4, 82–86.

18. Klaproth 1789a, 396.

19. Klaproth 1789b, 171.

20. Klaproth 1789a, 397; ABBAW, I-IV-32, 81.

21. Kopp 1843–1847, vol. 4, 82.

22. Klaproth 1789a, 399.

23. KPM Archiv XVII.12, 57. *Urangelb* (uranium yellow) is listed in a KPM table of pigments (from 1838) as pigment no. 28. See Köllmann and Jarchow 1987, 323.

24. Klaproth 1789a, 387. These reform efforts in mineralogy occurred in parallel with the reform of the chemical nomenclature connected with Lavoisier's name. On the reform of nomenclature, see Klein und Lefèvre 2007.

25. Based on the existing experimental techniques, it was impossible to decide whether uranium or, alternatively, uranium calx was a component of pitchblende, since it was possible that uranium calx resulted from the acidic dissolution of pitchblende.

26. Klaproth 1789a, 387.

Chapter 3

1. See Hufbauer 1982, 54–55; Klein 2007a; Klein 2007b; Spary 2010; Tomic 2010.

2. Quoted in Stürzbecher 1966, 49; the Brandenburg medical edict of 1693 is reprinted there.

3. This part of the discussion is based on Klein 2010c.

4. See Kopp 1843–1847, vol. 4, 299–302.

5. Frobenius 1729–1730, 285. Frobenius's remark on the harmony between ether and gold had alchemical connotations but was also based on the observation that ether attracted gold out of its dissolution with aqua regia (a mixture of nitric acid and hydrochloric acid).

6. Frobenius 1729–1730, 286.

7. Before 1741, the French chemists Etienne François Geoffroy and Jean Grosse had received samples of pure ether from Frobenius and Hanckwitz, and subsequently experimented on this substance. See Kopp 1843–1847, vol. 4, 303.

8. For these journals, see below. In France, the following publications on ethers were pathbreaking: Boullay 1807; Fourcroy and Vauquelin 1797; Macquer 1766, vol. 1.

9. The mixture of saltpeter and vitriolic acid yields nitric acid.

10. Göttling 1781a.

11. Göttling 1783b.

12. Göttling 1781b.

13. Göttling 1783a.

14. "Infusion" was the designation for materials extracted by mixing spirit of wine or purer forms of alcohol with a plant or animal substance.

15. Voigt 1784, 184.

16. Göttling 1778.

17. Göttling 1792.

18. See Hufbauer 1982, 207–208; Frercks 2008.

19. For this theory see Klein and Lefèvre 2007.

20. Macquer 1766, vol. 1, 461. This understanding was derived from the observation that ordinary ether was highly flammable, burned without a residue, and could be mixed with water—but to a lesser degree than spirit of wine.

21. Göttling 1779, 57.

22. Ibid., 58.

23. See Lehmann 1936.

24. On Henckel, see Herrmann 1962.

25. Stoeller 1800, 696. On Wiegleb, see also Hufbauer 1982, 190–191; Klosa 2009.

26. Hermbstaedt 1790, 94.

27. Ibid., 96.

28. Anonymous 1793a, 122; Anonymous 1793b; Trommsdorff 1793a, 80–81.

29. Trommsdorff, 1793b, II.

30. Trommsdorff, 1793c.

31. The name *Almanach* (almanac) refers to a calendar included at the beginning of the journal, which listed pharmaceutical preparations for each month.

32. Göttling 1786.

33. Ibid.

34. Göttling 1787.

35. Hufbauer 1982, 62; Brock 1993, 436.

36. Crell 1778, 12, 19. The journal's first title was *Chemisches Journal*. In 1781 it was renamed as *Die neuesten Entdeckungen in der Chemie*, and in 1784 it was renamed again as *Chemische Annalen für die Freunde der Naturlehre, Arzneygelahrtheit, Haushaltungskunst und Manufakturen*. For a more detailed analysis of Crell's journal see Hufbauer 1982, 62–95.

37. A list of all subscribers to Crell's journal in the period of 1784–1789 is published in Hufbauer 1982, 272–299.

38. The list was put together in preparation for war with Napoleon and the corresponding need to recruit field apothecaries. It is published in Adlung 1928.

39. This proportion of subscribers was distributed unequally between large and small towns. As practically all of Berlin's apothecaries were subscribers to Crell's journal, the Prussian capital was clearly a center of apothecaries' chemical activities.

Chapter 4

1. See Klein 2013; Klein 2014c; Kolbe 1863; Köllmann and Jarchow 1987; Siebeneicker 2002.

2. See Walcha 1975.

3. As early as 1763 Friedrich II ordered that orphans from the orphanage of Potsdam were to be employed in the KPM; Geheimes Staatsarchiv Preussischer Kulturbesitz (hereafter cited as GStA PK), 1 HA Geheimer Rat, Rep. 9 AV, E 166 II, Fasc. 3, 17, 21.

4. The following paragraph is based on Siebeneicker 2002.

5. Quoted in Siebeneicker 2002, 147.

6. Quoted in Köllmann and Jarchow 1987, 284.

7. Ibid.

8. See Klein 2014b.

9. Gerhard 1779.

10. Archive of the Königliche Porzellanmanufaktur (hereafter cited as KPM archive), XVII. 12 (Acta die Untersuchung des Farben-Laboratorii … betreffend), 8.

11. Quoted in Tetzlaff and Krohm 2007, 75. According to Tetzlaff and Krohm, the KPM is still using this recipe today and keeps it secret.

12. Lehman 2012, 331.

13. KPM archive, XVII. 12 (Acta die Untersuchung des Farben-Laboratorii ... betreffend), 9–10.

14. Ibid., 9.

15. Ibid., 10.

16. On this problem, see Klein 2014b.

17. KPM archive XVII. 12, 10.

18. KPM archive II 1, vol. 1 (Acta die Etablirung der Königlichen Porzellan Manufaktur-Commission ... betreffend), 28; KPM archive V 6 (Acta die Anstellung des Friedrich Bergling beim Laboratorio der Königlichen Porzellan Manufaktur betreffend), 2.

19. Ibid., 3.

20. Ibid., 10.

21. KPM archive XVII. 12, 16.

22. Ibid., 17.

23. Ibid., 24.

24. Ibid., 19.

25. Ibid., 20.

26. Klaproth 1788–1789.

27. KPM archive XVII. 12, 21.

28. Ibid., 11–14.

29. Ibid., 36–49.

30. Klaproth 1789d, 7.

31. KPM archive, XVII. 12, 50.

32. Ibid., 51–53.

33. Ibid., 54–56.

34. Ibid., 57–58 and 59.

35. Ibid. 58. The table was constructed by the KPM's director Georg Friedrich C. Frick; see Köllmann and Jarchow 1987, 323 (color number 28 of the table).

36. GStA PK, I. HA Rep. 151 Finanzministerium, Abt. IC, no. 9469, 64–71.

37. Siebeneicker 2002, 99.

38. KMP archive XVII. 12, 66–67.

39. Ibid., 68.

40. Ibid., 62.

41. Ibid., 65.

42. Ibid., 62.

43. Ibid., 70.

44. Ibid.

45. Ibid., 71.

46. Ibid., 72.

47. Ibid., 73.

48. Ibid., 60.

49. Ibid.

50. Heinitz initiated studies of improvements of furnaces as early as July 1787. See KPM archive XVII. 27, 18–21. See also Siebeneicker 2002.

51. KMP archive XVII. 12, 60.

52. GStA PK, I. HA, Rep. 151, Finanzministerium, Abt. IC, no. 9469, 33.

53. KPM archive XVII. 12., 76–79.

54. The name referred to the fact that it was not glazed with lead compounds, which were poisonous. See Siebeneicker 2002, 36–37.

55. Referring presumably to Roesch, Hermbstaedt wrote in a teaching report to the mining administration of October 13, 1796 that "one of the young arcanists of the Porcelain Manufactory" had attended his course. GStA PK, I. HA, Rep. 121, no. 7959 (Acta Gen. den wissenschaftlichen Unterricht der Berg-, Hütten- und Salinen-Aspiranten betreffed), 96.

56. Siebeneicker 2002, 149–150.

Chapter 5

1. See Krünitz 1773–1858, vol. 129, 421; Grimm and Grimm 1854–1956, vol. 8, 1610. In the late Middle Ages, the Latin term *expertus* had come into use, which denoted a person who possessed rare and accumulated experiential knowledge. The

English translation into *expert*, mostly used as an adjective, appeared shortly afterward, followed by the French substantive *l'expert*. The German term *Experte* appeared only in the nineteenth century, while in the centuries before Germans spoke of *Sachverständige* and *Sachkundige*; see Füssel 2012; Rexroth 2012.

2. See also Ash 2010a.

3. See also chapter 13.

4. Fleischer 2007.

5. But even in liberal Great Britain the state often encouraged, protected, and regulated such endeavors. See Ash 2007; Ashworth 2017; Jones 2008; Miller 2009; Tomory 2012; Tomory 2014.

6. Spary 2010.

7. Cook 2007; Dupré and Lüthy 2011.

8. Smith 2004.

9. Bennett 1987; Harkness 2007.

10. Dupré 2017.

11. Belhoste 2011; Beretta 2012; Beretta 2014; Kieffer 2014; Lehman 2012; Lowengard 2005; Nieto-Galan 2001; Hilaire-Pérez 2008.

12. See, for example, Pugliano 2012; Tomic 2010.

Chapter 6

1. See Baumgärtel 1963.

2. Agricola 1950 (1556), 99.

3. The term *Markscheider* is derived from *Markscheide*, the term for the borderline between mines, from the middle high German *marc, march* = (border)sign, borderland, and the old High German *marcha* = border, which is related to the Latin *margo, marginal*.

4. Sennewald 2002. On the designation of mine surveying as *"geometria subterranea"* and the mine surveyor as a "practical mathematician," see Baumgärtel 1965; Morel 2015; Wolff 1734, 842–844.

5. See Trebra 1785, v.

6. Agricola 1950 (1556), 96.

7. Baumgärtel and Wächtler 1965; Sennewald 2002. On the eighteenth-century published literature on assaying, surveying, mine construction, and mine machinery, see Baumgärtel 1965, 76–77, 93–103; Morel 2015.

8. Quoted in Schleiff 2013, 138.

9. See Baumgärtel 1965; Herrmann 1953; Schellhas 1959; Sennewald 1994; Schleiff 2013; Wagenbreth 1994.

10. Trebra recounts his experiences at the Freiberg Mining Academy in the early years in Trebra 1818. Alexander von Humboldt was in Freiberg from June 1791 to February 1792, around the same time as von Buch, while Novalis studied there from December 1797 to spring of 1799.

11. The *Advertissement* is reproduced in Oppel 1769, unpaginated.

12. Konečný 2013, 104. On education and training at other mining academies see Brianta 2000; Laboulais 2012; Schleiff und Konečný 2013; Weber 2015.

13. Humboldt 1973, 144. According to Schellhas, the Saxon scholarship recipients at the Freiberg Mining Academy were allocated hewing work in the mines, referred to as *Freigedinge*, for which they received a weekly wage. But Humboldt was not a scholarship recipient and here he was doing the work of an apprentice hewer. See Schellhas 1959, 355–356.

14. Humboldt 1973, 153–154. *Oryktognosie* was Werner's term for mineralogy.

15. Baumgärtel 1965, 138–141; Herrmann 1953; Weber 1976, 152–167; Schleiff 2013.

16. Zimmermann 1744; Zimmermann 1745; Zimmermann 1746; Justi quoted in Schleiff 2013, 129.

17. Zimmermann 1746, 54–55, 128. Zimmermann's conception of the mining sciences differed in some respects from those of Heinitz and the professors in Freiberg. For example, he had a positive attitude toward alchemy and dowsing rods.

18. Quoted in Herrmann 1953, 39–40. Heinitz's plan is published in Herrmann 1953.

19. Ibid., 39.

20. See Weber 1976, 146.

21. See Baumgärtel 1963, 80; Baumgärtel and Wächtler 1965; Schellhas 1985; Schleiff 2013; Sennewald 2002.

22. Trebra 1818, 9, 12.

23. The term *Geologie* (geology) only came into use at the end of the eighteenth century, while earlier the term *Geognosie* (geognosy), introduced by G. C. Füchsel and Werner, was commonly used. An exception is Alexander von Humboldt, who already used the term *Geologie* in his *Florae Fribergensis* (1793). In this work, Humboldt refers to "geographia oryctologica, which is simply called *geognosia* or geology, and which the brilliant Herr Werner has so wonderfully described." Quoted in Beck 1973, 168.

24. Werner 1791, 201.

25. See Fritscher 2013; Guntau 1984; Klein 2012c; Laudan 1987; Oldroyd 1996; Rudwick 2005. By contrast, in Hofbauer 2015, geology's mining context recedes into the background.

26. Werner 1787, 39. Werner's differentiation between minerals and types of rock or rock masses is set forth in this brief text, which today belongs to the classics of the history of geology. On this, see also Ospovat 1971; Werner 1791.

27. Only from 1786 onward did he use the term *Geognosie*. See Guntau 1984, 67.

28. Werner's unpublished papers, quoted in Guntau 1984, 79.

29. Ibid.

30. Bartels 2009, 329, 334. On the mine plans, see also Kroker 1972; Morel 2015; Sennewald 2002; Wagenbreth 1996.

31. See Guntau 1984, 69–89; Gohau 1990; Laudan 1987, 87–112.

32. Werner initially classified basalt as *Urgebirge*, but later assumed that it was formed in a later geological era by erosion and subsequent sedimentation in water. The competing vulcanist theory postulated, by contrast, that basalt, granite, gneiss, and other crystalline rocks were the products of volcanic activity.

33. In his lectures after 1790, Werner added a third type of rock—the *Übergangsgebirge*—or transitional rock. The volcanic rocks played only a marginal role in Werner's classification.

34. Werner 1791, dedication; xiii. See also Werner 1809.

35. The argument that it was mere rhetoric is presented in Dym 2011; Vogel 2013; Wakefield 2009.

36. On this, see also Guntau 1984, 67–68, 86–97.

37. On the long-term nature of such innovation strategies, see also Popplow 2010b.

38. Agricola 1950 (1556), 38–41. By contrast, Dym and Vogel take the rejection of the dowsing rod in the Enlightenment as evidence that the scientifically educated mining officials rejected the practical knowledge per se; Dym 2011; Vogel 2013.

39. Agricola 1950 (1556), 39, 41. The occasional successes of the dowsers can be explained by the fact that they often looked for deposits in areas where there were closed mines, and that they were knowledgeable of the other characteristics mentioned by Agricola.

40. Friedrich Wilhelm von Trebra, mining master and educated at the Freiberg Mining Academy, tolerated the use of the dowsing rod in the area under his responsibility, although he personally was opposed to it. See Trebra 1818.

41. Guntau 1984, 90.

42. See Zimmermann 1744, unpaginated; Zimmermann 1746, 54–55, 128.

Chapter 7

1. On Lehmann see Freyberg 1955.

2. On Hagen's reform project, see Johnson 1975.

3. Such a double appointment had already been made in the case of the mineralogist Johann Gottlob Lehmann.

4. Up to his retirement in 1810, his salary rose to the remarkable sum of 3,286 thalers. In addition, he received an annual pension of 600 Reichsthalers from the Academy of Sciences; Wutke 1913, 437, 441.

5. One result of this work was the production in 1782 of the first gun barrels of Silesian iron, under Gerhard's supervision; this took place in the gun manufactory at Spandau. See Wutke 1913, 439.

6. Gerhard 1773–1776, vol. 1, introduction.

7. On Reichardt see Johnson 1975

8. See Fechner 1900–1901, vol. 48, 310.

9. Treue 1984, 57–60, 88–89.

10. See Fechner 1900–1901, vol. 49, 55.

11. The most comprehensive contemporary exposition of iron and steel production, which appeared shortly after Gerhard's first inspection tour, was Antoine-Gabriel Jars's four-volume *Voyages Métallurgiques* (1774–1781). Years later, Gerhard translated this book with annotations into German; Gerhard 1777–1785.

12. Quoted in Fechner 1900–1901, vol. 49, 35.

13. Ibid., 36–37.

14. Lehmann 1752.

15. Ibid., 31. On Lehmann's approach, see Oldroyd 1996; Laudan 1987.

16. Fechner 1900–1901, vol. 49, 37–44.

17. Ibid., 21–28, 38–39. Saxony was rich in cobalt ore and exported Saxon blue to England, the Netherlands, and many German states. The manufacture of Saxon blue was secret and strictly regulated by the state. The Saxon state further prohibited the export of raw cobalt ore.

18. In 1768, Justi was accused of fraud and jailed; see Wakefield 2009.

19. Wutke 1913, 35, 39, 439.

20. This report is no longer extant, but excerpts from it are quoted in Fechner 1900–1901, vol. 49, 249.

21. The Academy secretary reported in the customary French: "M. Gerhard a présenté des minéraux qu'il a rapportés de son dernier voyage de Schlésie." See the *Registres* of the Akademie der Wissenschaften, August 21, 1766, to August 17, 1876, ABBAW, I-IV-32, 67.

22. Ibid., 68 ("mémoir systématique sur les Minéraux"), 70 ("M. Gerhard a lu des Remarques physiques & minéralogiques sur les montagnes de Schlésie").

23. Gerhard 1771.

24. Agricola 1950 (1556), 30.

25. See also Humboldt's reports analyzed in chapter 10.

26. Gerhard 1771, 100.

27. Ibid., 110.

28. Ibid., 104–105. On the classification of types of mountains, related to this discussion, see also Gohau 1990, 79–81.

29. The concept of "physical geography" was introduced in Gerhard 1781–1782, vol. 1, 25.

30. He reported among other things that Count von Schaffgotsch had begun to set up a tin mine before the Seven Years' War but then had the misfortune to have turned to "persons of inferior intellegence"—by which he meant the responsible officials—and therefore had to abandon his undertaking; Gerhard 1771, 105.

31. Werner 1787; Werner 1971; Gerhard 1781–1782, vol. 1. Werner used the term *Mineralogie* later in the same broad sense as Gerhard. See Ospovat 1971, 101.

32. An example for this is a review by G. H. I., published in the *Leipziger Magazin zur Naturkunde, Mathematik und Oekonomie* (1781), 104–115, 581–528.

33. Rudwick 2005, 84. See also Guntau 1984; Laudan 1987; Oldroyd 1996.

34. GStA PK, I. HA, Rep. 121, no. 7958, 95.

Chapter 8

1. GStA PK, Rep. 121, no. 141, 5, 11, 13, 14, 24.

2. GStA PK, I. HA, Rep. 121, no. 7958, 53.

3. GStA PK, I. HA, Rep. 121, no. 141, 14–24; GStA PK, I. HA, Rep. 121, no. 7958, 55–60.

4. Ibid., 64–69.

5. Kopp 1843–1847, vol. 4, 150–155.

6. Gerhard 1779, 13.

7. Gerhard 1780.

8. In 1782, the first production of gun barrels from Silesian iron took place under his supervision, and shortly thereafter he participated in trials of cannon casting in Malapane, in Silesia. See Fechner 1900–1901, part 2, 437–440; Wutke 1913, 439.

9. Gerhard 1781a, 102.

10. Gerhard 1781b, 117, 121, 123.

11. GStA PK, I. HA, Rep. 121, no. 7958, 144.

12. Ibid.

13. Ibid., 145.

14. Ibid., 146.

15. Autographensammlung der Bibliothek der Humboldt Universität no. 411, 1.

16. In the same year, Gerhard published another textbook on mineralogy; see Gerhard 1786.

17. Gerhard 1797.

Chapter 9

1. On parallel developments in other countries, see the overviews in Brianta 2000; Schleiff and Konečný 2013; Weber 2015. For specialized studies, see: on Austria-Hungary, Konečný 2013; on the Harz, Bartels 1992, 384–388; Bartels 2013; on France, Laboulais 2012.

2. Krusch 1904. Krusch's assertion has been disproven in Klein 2010a; subsequently as well in Engel 2013.

3. GStA PK, I. HA, Rep. 121, no. 7957, 9.

4. Ibid., 15–16.

5. Ibid., 28, 30.

6. Ibid., 33.

7. Preußische Akademie der Wissenschaften 1892–1982, vol. 15, 280; GStA PK, I. HA, Rep. 121, no. 7957, 16.

8. Ibid., 42–43.

9. Ibid., 34–35.

10. Ibid., 111; italics mine.

11. Ibid., 122; italics mine.

12. This corresponded to the salary of a *Sekretär*, a middle-ranking official.

13. GStA PK, I. HA, Rep. 121, no. 7957, 110, 146, 160. Rose died on April 28, 1771.

14. Ibid., 162, 172.

15. Ibid., 165–166; GStA PK, I. HA, Rep. 121, no. 7958, 2–4.

16. Ibid., 53. In the previous years, his teaching had included only some modest experimentation, carried out with a portable assaying furnace and an assaying balance he had purchased in January 1771; see GStA PK, I. HA, Rep. 121, no. 141, 4.

17. Gerhard 1773–1776, vol. 1, 13.

18. Gerhard 1773–1776, vol. 2, *Vorrede*, n.p.

19. Radkau 1989, 59–73.

20. Wutke 1913, 37.

21. In Walter's case, it turned out that his courses were identical with his physical course at the Medical-Surgical College, to which he apparently invited the future officials. GStA PK, I. HA, Rep. 121, no. 7958, 46–47.

22. GStA PK, I. HA, Rep. 121, no. 7957, 86.

23. Ibid., 87.

24. GStA PK I. HA, Rep. 121 Ministerium für Handel und Gewerbe, Abt. Bergwerks-, Hütten- und Salinenwesen no. 277, 9–10.

25. See chapter 4. On Bückling, see Weber 1976; on Eversmann, see Breil 1977.

Chapter 10

1. Humboldt 1973, 137. For Humboldt's early career see also Klein 2012a; Klein 2015.

2. For the young Humboldt's studies of plants see Anthony 2018.

3. Quoted after Bruhns 1969, vol. 1, 134–135.

4. Humboldt 1973, 175.

5. Quoted in Köhl 1913, 105.

6. Humboldt 1973, 218.

7. Humboldt 1959 (1792), 75.

8. Ibid.

9. Ibid.

10. Ibid., 101–103.

11. Ibid., 139.

12. Ibid., 140.

13. Ibid., 142.

14. Ibid., 145.

15. Humboldt 1973, 209.

16. Ibid., 211.

17. Ibid., 251.

18. Humboldt 1959 (1792), 124–131.

19. Ibid., 124, 127.

20. Humboldt 1973, 310.

21. Teich 1975.

22. Humboldt 1973, 275.

23. Ibid., 258, 265, 312.

24. Ibid., 344, 346.

25. Ibid., 352.

26. Ibid., 361–364.

27. Ibid., 378.

28. Quoted in Bruhns 1969, vol. 1, 164.

29. Humboldt 1973, 487–488.

30. Humboldt 1796.

31. Humboldt 1795, 107 and 109 (footnote).

32. Ibid., 106.

33. Humboldt 1796, 100.

34. Humboldt 1973, 504.

35. Ibid., 533.

36. Ibid.

37. Humboldt 1796.

38. Humboldt 1973, 524.

39. Ibid., 534.

40. Ibid., 524, 529.

41. Humboldt 1799, iv.

42. A recent example is Wulf 2015.

43. For more details see Klein 2015.

44. Humboldt 1797, 2, 14. Humboldt's report was written by a clerk and signed by Humboldt. It is preceded by a letter by Humboldt of February 24, 1797, in which he states that the original report was revised three times on behalf of Hardenberg, who required the data for his own statements of accounts. Supplementing the main report, there are also twenty-one notes by Humboldt, written between 1795 and 1797.

45. Humboldt 1973, 311.

46. Humboldt 1797, 18.

47. Ibid., 6.

48. Ibid., 6.

49. Quoted in Beck 1959, vol. 1, 50.

Chapter 11

1. Preface to the German edition (1557), in Agricola 1977 (1557), xxviii.

2. Oppel's *Bericht vom Bergbau* was originally written by the Saxon gemstone inspector Johann Gottlieb Kern; Oppel published the work as a textbook. See Baumgärtel 1965, 146–148; Weber 2015, 237–240, 247–252.

3. Oppel 1769, *Vorbericht*, n.p. In the preface Oppel states further that the book serves as a teaching manual (*Leitfaden*) for the oral instruction of some young people "in the miners' sciences [*in Bergmännischen Wissenschaften*]."

4. Delius 1773, *Vorbericht*, n.p.

5. Peithner 1769–1770, *Vorrede*, n.p. On this, see also Fessner 2005, 771–772; Weber 2015, 240. The chair of mining science at the University of Prague was established in 1762.

6. Heinitz 1963 (1771).

7. Ibid., 163.

8. Agricola 1950 (1556), 30–32. Zimmermann's and Heinitz's criticisms also reveal the discontinuous historical impact of early technological texts on mining.

9. Zimmermann 1744, 128.

10. Charpentier 1778. Petrographical or geological maps are topographical maps that record the rock formations. On the geological mapping of Saxony, see Freyer 1985.

11. Heinitz 1963 (1771), 132, 136.

12. Ibid., 139, 136.

13. Ibid., 168.

14. Ibid.

15. On the concept of "reliable" useful knowledge, see also Inkster 2006.

16. Heinitz 1963 (1771), 168.

17. GStA PK, I. HA, Rep. 121, no. 7957, 2–6.

18. Ibid., 2.

19. GStA PK I. HA, Rep. 121, no. 7958, 130–135.

20. Ibid., 131.

21. See the overview in Harwood 2010.

22. Konečný 2013.

23. GStA PK, I. HA, Rep. 121, no. 7957, 2.

24. Ibid., 2–3.

25. Ibid., 2; italics mine.

26. See Klein 1994.

27. GSTA PK, I. HA, Rep. 121, no. 7957, 3.

28. See chapter 6.

29. GStA PK I. HA, Rep. 121, no. 7957, 2.

30. Laboulais 2012, 16, 287.

31. Quoted in Laboulais 2012, 221.

32. Bartels 1992, 372.

33. Ibid., 379; italics mine. See Calvör 1763.

34. Baumgärtel 1965, 150.

35. Vogel 2013, 20. See also Dym 2011; Wakefield 2009.

36. Weber 2015, 219.

Chapter 12

1. Vogel 2008.

2. Humboldt also discussed briefly the question of improving saltworks, based on the contemporary chemical and physical theories of vaporization and evaporation. However, he considered these theories too abstract to be linked to technological knowledge; Humboldt 1792, 17–22, 117–121.

3. Humboldt 1792, 24–44, 97–118.

4. Ibid., 2. On Humboldt's visits to the saltworks, see Biermann, Jahn, and Lange 1968, 3, 7.

5. The argument that it was wholesale replacement has been claimed repeatedly in the literature. See, for example, Vogel 2008.

6. Humboldt 1792, 140.

7. Ibid., 23, 39.

8. See Humboldt 1973, 61, 70.

9. Gmelin 1786, iii–iv.

10. Humboldt 1792, 3, 2.

11. See Klein 2015, 68–69.

12. Common salt, sodium chloride, could not be directly decomposed into sodium and chlorine with the chemical methods of the time. The decomposition involved instead the formation of two new compounds, *sel alcali fixe* (sodium carbonate) and muriatic acid (hydrochloric acid), which contained the two components sodium and chlorine.

13. In this two-step process, common salt was first decomposed with sulfuric acid and transformed into sodium sulfate. Sodium sulfate was then converted into soda (sodium carbonate) through the addition of coal (carbon) and limestone (calcium carbonate). On the historical development of the Leblanc process, see Osteroth 1985; Smith 1979, 192–306. On the relationship between chemistry and the Leblanc process see Gillispie 1957b.

14. Humboldt 1792, 14–15.

15. Ibid., 5–6.

16. Ibid., 6.

17. Ibid., 10.

18. Ibid., 16. On the lead chamber method and its roots in chemistry, see Osteroth 1985; Smith 1979; on Born's amalgamation, see Teich 1975.

19. On the system of concepts connected to this, such as chemical compound, composition, and analysis, see Klein 1994.

20. Humboldt 1792, 16–17.

21. Ibid., 139–140.

22. Radder writes, "Technology is taken to embrace the technological sciences, while the technological sciences include several disciplines in addition to the engineering sciences, such as information science, medical science, and agricultural science. Making such a direct link between technology, more broadly, and the technological sciences makes sense in view of the fact that these sciences aim to contribute towards realizing contemporary or future technologies." Radder 2009, 65.

23. Meijers 2009. In three contributions to the same volume, engineering science is treated as a distinct issue: see Banse and Grunwald 2009; Boon and Knuuttila 2009; Channell 2009. For a discussion of technological science from a philosophical perspective, see also Hansson 2007.

24. The term "experimental history" was introduced by Francis Bacon, who outlined his ideas of an experimental history in *Preparative towards a Natural and Experimental History* (1620), published in the same year as his *Novum Organum*. For Bacon's concept of experimental history see Klein and Lefèvre 2007. For his concept of experiment see also Klein 1996.

25. For the writing of his *De re metallica*, Agricola drew on the technological literature, namely, the assayers' booklets; see Baumgärtel 1965; Darmstaedter 1926. On the shift from local technical knowledge to technological texts see also Popplow 1998; Popplow 2015.

26. When early eighteenth-century chemists investigated affinities between chemical substances by exploring displacement reactions, for example, it was irrelevant in their view whether these reactions occurred in nature or in the laboratory, that is, in chemical art. See Klein 1994; Klein and Lefèvre 2007.

27. Bacon 1986–1994, vol. 4, 47–48.

28. Ibid., 257–258.

29. On the economic and patriotic societies in the German states, see Popplow 2010a; Popplow 2010b.

30. On cameralism and cameral science, see Brückner 1977; Dittrich 1974; Lindenfeld 1997; Olesko 2020; Small 1909; Sokoll 2007; Troitzsch 1966; Wakefield 2009.

31. Beckmann mentioned only in passing that technology had also the task to "explain the phenomena that occur in processing." Beckmann 1777, xv. On

Beckmann, see Bayerl and Beckmann 1999; Banse and Müller 2001. On the meaning of technology (*Technologie*), see also Schatzberg 2018; Timm 1964.

32. See Ropohl 2006.

Chapter 13

1. See Klein 2012a; Klein 2012b; Klein 2012c; Klein 2017. See also chapter 17.

2. On these persons' administrative knowledge see also Felten 2018.

3. For an overview of the literature, see the introduction.

4. Long 2011, 62; italics mine.

5. Chalmers 2017, 27.

6. Damerow and Renn 2010. See also Lefèvre 2001; Valleriani 2010.

7. Schemmel 2008, 15. For similar figures see also the essays included in Cormack, Walton, and Schuster 2017.

8. See chapter 3.

9. Smith 2004. Many additional persons Smith discusses in her book are, in my view, scientific-technological experts as well; Smith classifies them as artisans.

10. Jones 2008.

11. See Kocka 1969; Wise 2018.

12. On the Imperial Institute of Physics and Technology, see Cahan 1989; Cahan 2018.

13. Layton 1971, 578.

14. Merton 1970 (1938); Merton had published his article in 1938 in the journal *Osiris*. For Zilsel, see the edition of his texts in Zilsel 2000; for Hessen and Grossmann, see the edition of their texts in Freudenthal and McLaughlin 2009. On the scholar-and-craftsman thesis see also Cohen 1994.

15. Merton 1970 (1938), 146.

16. Hall 1959, 17. Hall criticized Merton's thesis in the context of a conference which took place in 1957 at the University of Wisconsin.

17. Hall 1963. Henceforth, what was regarded as the "Merton thesis" included only those parts of his argument that addressed the influence of the Protestant ethic on the early modern sciences. On the political background of Hall's criticism, see also Klein 2016b. The "scholar-and-craftsman thesis" has sometimes also been ascribed to Edgar Zilsel; see Zilsel 2000.

18. Krige 2006, 11.

19. Gillispie 1981b, 581.

20. Gillispie 1980.

21. Gillispie 1957a, 402, 403.

22. Gillispie 1980, 86–87.

23. Ibid., 551. In an earlier passage on Berthollet, he wrote, "It was definitely the opportunity in applied chemistry that drew Berthollet, even like Darcet before him, from the service of the duc d'Orléans into that of government, and thereby allowed him to make the transition from a livelihood in medicine to one in science." Ibid., 409.

24. Further evidence of this is cited in Klein 2015.

25. Gillispie 1980, 409.

26. Shapin 1994, 361.

27. Merton 1970 (1938), 146.

28. Dear 2007, 436.

29. Ibid., 435.

30. See, for example, Belhoste 2012; Beretta 2012; Beretta 2014; Lehman 2012; Spary 2013.

31. Hahn wrote: "The same objections [as those against scholars], however, could not be held out against the craftsman, who displayed, as D'Alembert pointed out in the *Encyclopédie*, 'the most admirable evidences of the sagacity, the patience, and the resource of the mind' that intellectuals envied. The new seeker of knowledge and the artisan ought to have become brothers-in-arms." Hahn 1971, 41.

32. Ibid.

Chapter 14

1. See Müller-Wille and Böhme 2012; Wise 2018. By contrast, the royal art collections were not moved to the university building; see Bredekamp and Labuda 2012, 240.

2. See, for example, Tenorth 2012; Turner 1971; Smith 1991; Wise 2018.

3. Lenz 1910–1918, vol. 1, 66. See also Tress 2010.

4. Quoted in Tenorth 2012, xxiv.

5. In 1808, the General Directory was reorganized and renamed the State Ministry (Staatsministerium).

6. Quoted in Treue 1984, 19. On the establishment of the professorship for *Technologie*, see Timm 1960, 48. On the four Prussian universities, see Clark 2006, 52.

7. See Kleinert 2013.

8. GStA PK, I. HA, Rep. 121, no. 7957, 10.

9. Ibid., 82–92 (directive for the philosophical faculty); 93–100 (directive for the law faculty); 101–102 (directive for the medical faculty); 103–106 (directive for the theological faculty). On November 22, 1770, these directives for all four faculties were approved by the king and sent to all the chambers and departments (ibid., 152).

10. Ibid., 18–20, 58.

11. Ibid., 50–53, also with extracts from the course catalog.

12. Ibid., 153–157. There are no related documents about the University of Duisburg.

13. Ibid., 21, 25–25.

14. Ibid., 175.

15. GStA PK, I. HA, Rep. 121, no. 7958, 27, 28.

16. GStA PK, I. HA, Rep. 121, no. 7957, 22.

17. Quoted in Dann 1958, 47.

18. GStA PK, I. HA, Rep 151, IC, no. 8746 Acta des Finanzministerii betr. Die Ernennung des Obermedizinalraths Klaproth zum Professor der Chemie, 1810, p. 2.

19. See chapters 1 and 2.

20. This additional private teaching was arranged by Gottlob Johann Christian Kunth, who was the main teacher of Wilhelm and Alexander von Humboldt. Later, Fischer was also a private teacher of Moses Mendelssohn's sons as well as of Crown Prince Friedrich Wilhelm and Prince Wilhelm (later Kaiser Wilhelm I); see Klemm 1966.

21. GStA PK, I. HA, Rep. 121, no. 7959, 116.

22. Ibid., 146–147.

23. Ibid., 148.

24. Ibid., 128, 132.

25. Ibid., 137.

26. Ibid., 200.

27. See Klein 2010b.

28. Klemm and Meyer 1968, 90.

29. Lenz 1910–1918, vol. 1, 257–258.

30. Ibid., vol. 2, 251.

31. Wise 2018, xvi. See also Wise's list of university professors who were teachers of Helmholtz; ibid., 13.

Chapter 15

1. It should be noted that the term "civil engineering" is a free translation into English of the German terms *Bauen* (building) and *Baukunst* (art of building, including *Zivilbaukunst*, *Oekonomie-Baukunst*, and *Wasserbaukunst*). By contrast, the terms General Building Department (*Oberbaudepartment*), building concillor (*Baurat*), and building official (*Baubeamter*) are literal translations.

2. Quoted in Strecke 2000, 121–122.

3. See Olesko 2009.

4. Kahlow 2000, 36–37; Simon 1902, 677–679; Strecke 2000, 122–124.

5. Strecke 2000, 119.

6. GStA PK, I. HA, Rep. 121, no. 7957, 41. On the Academy of Arts, see also Hannesen 2005; Müller 1896; Simon 1902.

7. Lundgreen 1975, 12–16; Simon 1902, 655–680; Strecke 2000, 124–125.

8. This context is detailed in Olesko 2009; see also Jaeckel 2000; Kahlow 2000; Vogler and Vetter 1973.

9. The school's name was Lehranstalt zum Unterricht junger Leute in der Land- und Wasserbaukunst. See Olesko 2009.

10. See also Kahlow 2000, 42; Olesko 2009; Strecke 2000, 44–46.

11. Quoted in Strecke 2000, 133.

12. Quoted in Strecke 2000, 132.

13. Quoted in Strecke 2000, 134.

14. See Dobbert 1899, 24; Kahlow 2000, 44; Olesko 2009, 26–27

15. Olesko 2009, 26.

16. Dobbert 1899, 26.

17. Ibid., 35.

18. Ibid., 26–27.

19. Wefeld 2000; Dobbert 1899, 24–25.

20. By contrast, Lundgreen argues that the Bauakademie imitated the *Écoles d'application*, which would imply that theoretical education played a minor role there (Lundgreen 1975, 32). Wise, who regards the École des Ponts et Chaussées as the model for the Bauakademie, also assumes that there were obstacles of principle to linking theoretical education with practical training; he thus asserts that the young Bauakademie "had difficulty meeting the demand for theoretically sophisticated architects and engineers who also possessed 'positive,' or practical, knowledge" (Wise 2018, 43).

21. See also chapters 3 and 14. For the combination of theory and practice at the War School and the United Artillery and Engineering School, see Wise 2018, 105–143.

22. These teachers did not receive an additional salary for their teaching. A salary for all teachers was introduced only after 1848. See Wefeld 2000.

23. See Radkau and Schäfer 1987; Strecke 2000, 86–91.

24. Eytelwein 1801.

25. The journal appeared until the beginning of the war in 1806. See Strecke 2000, 93; Kahlow 2000.

26. Gilly 1797–1798.

27. Quoted in Strecke 2000, 99.

28. Strecke 2000, 102.

29. In the meantime, the Bauakademie had moved into another building, located at the corner of Zimmerstraße and Charlottenstraße.

30. Wefeld 2000. The unification of the Academy of Arts with the University of Berlin had been proposed by Wilhelm von Humboldt; see Tenorth 2012, 23; Bredekamp and Labuda 2012, 239.

31. The documents about this period are incomplete; see Wefeld 2000.

32. In 1804 the General Building Department (Oberbaudepartment) was reorganized and renamed the Oberbaudeputation. Another reorganization of ministries took place in 1817, which led to the integration of the Building Department into the Ministry of Trade as its Abteilung für Bauwesen. In the following I keep the name Building Department.

33. Quoted in Lundgreen 1975, 33. See also Wise 2018, 44–45. In 1817 both an independent Cultural Ministry (Minsterium der Geistlichen, Unterrichts- und Medinzinalangelegenheiten) and a Ministry of Trade were established. The latter was subdivided into two sections, the Building Department (Abteilung für Bauwesen) and the Department of Trade and Industry (Abteilung für Handel und Gewerbe). The independent Ministry of Trade, headed by Hans von Bülow, existed only until 1825. From 1825 until 1848 it again became a department either of the Interior or the Finance Ministry. See also Mieck 1965b, 30.

34. For further details concerning related rivalries between the Ministries of Trade and of Culture, see Lundgreen 1975, 35–39; Wise 2018, 44–46.

35. Quoted in Dobbert 1899.

36. Wefeld 2000.

37. In the following I keep the name Bauakademie. On Beuth, see also chapter 15.

38. Dobbert 1899; Wefeld 2000.

39. Kathryn Olesko, personal communication.

40. In the 1840s Beuth began to open its doors to future private master builders as well.

41. Dobbert 1899, 48. See also Wise's translation of the classroom curriculum (after Dobbert) in Wise 2018, 49–50. Wise discusses classroom teaching from four points of view: aesthetics, drawing and construction, mathematics and mechanics, and experimental and natural-historical sciences; ibid., 48–55. As Wise points out, his grouping is informed by relating the curriculum to museums of art and natural history collections. By contrast, my own grouping into mathematical, natural-scientific, and technological instruction is informed by relating the Bauakademie to mining academies, schools of agriculture and forestry, and the Industrial Institute, that is, to the new institutions elaborating and teaching the useful sciences.

42. Quoted in Dobbert 1899, 54.

43. For the following see Dobbert 1899, 55–68.

Chapter 16

1. For further details see Klein 2015, 36–48, 237–241. The name of Struensee's department was the Fabriken-, Handels- und Akzisendepartment.

2. Quoted in Mieck 1965a, 344.

3. It was a formal part of the administration, got an annual budget, and its members received a modest salary.

4. Quoted in Matschoß 1911, 243.

5. See chapter 12.

6. Quoted in Mieck 1965b, 33.

7. Quoted in Mieck 1965b, 96.

8. Goldschmidt and Goldschmidt 1881, 32–34; Mieck 1965b, 101–107.

9. Matschoß 1911, 255.

10. See chapter 14.

11. Mieck 1965b; Straube 1930; Wise 2018, 45–48, 68–99.

12. Henderson 1958, 113–114; Mieck 1965b, 90–92; Wise 2018, 84–85.

13. Mieck 1965b, 96–97.

14. Lundgreen provides a number of examples for this practice; see Lundgreen 1975, 178–190.

15. Quoted in Matschoß 1911, 239.

16. Quoted in Matschoß 1921, 119.

17. Mieck 1965b, 35–37; Wise 2018, 86–92.

18. For a more complete description of these plans see Matschoß 1911, 251–255.

19. Beuth 1822, 133–134.

20. See chapter 8.

21. Quoted in Mieck 1965b, 71–72.

22. Beuth 1822, 137.

23. See Mieck 1965b, 78–87; Matschoß 1911, 263–266. The Prussian patent law was established in October 1815; a central institution for patents (Reichspatentamt) was founded in 1877.

24. On the relationship between mercantilism and liberalism in this period see Mieck 1965b.

25. This does not mean that useful science promoted by the different state departments had exactly the same theoretical level. For example, advanced mathematics played no role at the Industrial Institute.

26. See Beuth's plan from April 1821, reprinted in Nottebohm 1871, 3–8.

27. Beuth 1822. In Prussia, the traditional apprenticeship started at the age of fourteen. This tradition was not immediately abolished after the introduction of freedom of trade in 1810; see Lundgreen 1975, 138, 144.

28. After the abolition of guild compulsion and introduction of freedom of trade in 1810, the building sector long remained regulated by the state for reasons of safety; this implied continuation of training via apprenticeship with an established master. See Lundgreen 1975, 143–165.

29. See also Wise 2018, 77–86, especially his translation of the curriculum in 1826; ibid., 83.

30. Beuth 1822, 142.

31. Lundgreen 1975, 116, 142–165, 178–272; Lundgreen 1990, 45; König 1998, 120; König 2006a, 194. Lundgreen and König have argued that other technical schools in the German states were significantly less successful in terms of industrial positions of their students.

32. Beuth 1822, 142.

33. For the occupations of *Werkmeister* and *Werkstattvorsteher* see Kocka 1969, 72–76; 93–101.

34. Lundgreen 1975, 143–165.

35. By contrast, as with the foundation of the Bauakademie (see chapter 14), Lundgreen puts the foundation of the Industrial Institute in the context of the French system of École Polytechnique and *Écoles d'application*; Lundgreen 1975, 30–32. In a similar way, Wise argues that the Conservatoire National des Arts et Métiers in Paris "provided an instructive example" for the Industrial Institute; Wise 2018, 77.

36. See chapter 14.

37. Nottebohm 1871, 70–71; Beuth 1822.

38. Quoted in Cole 1951, 130.

39. The following paragraph is based on Tomory 2012.

40. Cole 1951, 136.

41. Quoted in Matschoß 1921, 149. See also Kocka 1969, 72.

42. Quoted in Lundgreen 1975, 143, 42.

43. See also Kocka 1969, 166–171; König 2000.

44. Weber 1854, 99–101. Weber designated himself an "engineer." In his essay, he used both the terms *Ingenieur* and *Techniker*. He designates the group of *Techniker* as a new "class of citizens" (*Klasse von Staatsbürgern*), but speaks more often of the "*Stand der Techniker*," here translated with "class" of *Techniker*. Weber 1854, 99.

45. Weber 1854, 105–106; Kocka 1969, 167–168. By contrast, Wise puts the Industrial Institute and the Academy of Civil Architecture into the context of the *Bildungsbürgertum*; see Wise 2018. It should be noted that the term *Bildungsbürgertum* was introduced in the early twentieth century. It served not least the goal of demarcating the humanistically educated elite from the new, competing elite of academically trained scientific experts and engineers. As a matter of fact, the use of the term *Bildungsbürgertum* has never been extended to all educated middle-class men and women; it has rather served to actively exclude engineers and industrialists as well as specialized empirical scientists.

46. This issue is detailed in Lundgreen 1975; see, in particular, his summarizing table (ibid., 190) and his statement about changes over time (ibid., 140).

47. Nottebohm 1871, 25. Nottebohm was the director of the Industrial Institute from 1857 to 1868.

48. Nottebohm 1871, 29, 39–40.

49. Grashof 1864.

50. The German term *Technische Hochschule* is usually translated as "technical university." "Technological" university would be more appropriate, since it was a place of *academic knowledge* about technology (*Technik*).

51. Quoted in Lundgreen 1975, 76.

52. Nottebohm 1871, 50.

53. Grashof 1864, 594, 597, 601. Twelve years later, Grashof argued again for the unification of the Bauakademie and the Gewerbeakademie in Berlin; see Grashof 1876.

54. Grashof 1864, 592, 604, 607, 609. See also Grashof 1876, 627.

55. Grashof 1864, 603.

Chapter 17

1. On controversies about more "practice-oriented" and more "science-oriented" engineering education in the late nineteenth and early twentieth centuries, see Harwood 2005; Harwood 2006; Harwood 2010.

2. Calvör 1763, quoted from the title of this book; Bartels 1992.

3. Layton 1971, 562.

4. See Alder 1997; Belhoste 2003; Belhoste 2011; Braun 1977; Bret 2002; Buchheim and Sonnemann 1990; Hilaire-Pérez 2000; Hilaire-Pérez and Verna 2006; Kahlow 2000; König 1981; König 2006b; Langins 2004; Lundgreen 1975; Picon 1992; Popplow 2006; Popplow 2015; Meyer 2012; Ropohl 1997; Ropohl 2006; Vérin 1993.

5. In his essay Layton used both the terms "engineering science" and "technological science," as well as the term "technology" in the sense of a science. For the history of the discussion fueled by Layton, see Channell 1989; Channell 2009. For a discussion of technological science, see also Banse, Grunwald, König, and Ropohl 2006; Banse and Grunwald 2009; Buchheim and Sonnemann 1990; Boon and Knuuttila 2009; Hansson 2007; König 2006b; Radder 2009; Wengenroth 2006.

6. Layton 1971, 567, 572.

7. Ibid., 568.

8. Ibid., 569.

9. Ibid., 575.

10. Ibid., 578.

11. Channell 1989, xxv. For the incentives Layton's model provided to subsequent studies of the technological sciences and their relationship with natural sciences, see Channell 1989, xxv–xxxi; Staudenmaier 1985.

12. König 1995, 3–4.

13. Ibid., 324–359.

14. On the formation of scientific disciplines see Henschel 2011; Laitko 2002; Stichweh 1992.

15. Landecker 2007.

16. Here Rheinberger's distinction between technical and epistemic objects is relevant; see Rheinberger 1997.

17. Landecker 2007, 70.

18. The term "biotechnology" is often restricted to later enterprises beginning in the 1970s, such as studies of recombinant DNA and their industrial application and commercialization. See, for example, Hughes 2001.

19. Landecker 2007, 17.

20. For examples of the ways in which discovery and invention converged in nineteenth-century organic chemical synthesis, see Jackson 2014.

21. Latour 1987, 1–2 (see his examples). The term "technoscience" was used before Latour in French philosophical discourse; see Hottois 2018; Lyotard 1984. On technoscience, see also Barnes 2005; Bensaude-Vincent and Loeve 2018; Channell 2017; Forman 2007; Houkes 2009; Hottois 2018; Ihde and Selinger 2003; Klein 2005a; Klein 2005b; Lefèvre 2005; Nordmann, Radder, and Schiemann 2011; Pickstone 2000; Rocke 2018.

22. For similar examples, see Rocke 2018. The terms "pure science" and "basic research" are not synonymous. Whereas in the late nineteenth and early twentieth centuries the concept of pure science was used to preserve science's independence from any utilitarian goals, the concept of basic research, introduced around the middle of the twentieth century, is linked to the argument that autonomous research might eventually pay off. For this distinction see Clarke 2010; Schauz and Kaldeway 2018, 1.

23. Layton 1974, 31. See also Barnes 2005; Schatzberg 2018.

24. Channell 2017, 163.

25. Weber 2011, 160–161.

26. For this argument see Nordmann 2011; Forman 2007.

27. For the concept of Anthropocene see Renn 2020.

REFERENCES

Archival Sources

ABBAW Archiv der Berlin-Brandenburgischen Akademie der Wissenschaften (Archive of the Berlin-Brandenburg Academy of Sciences and Humanities).

Autographensammlung der Bibliothek der Humboldt Universität zu Berlin (Autograph Collection of the Library of Humboldt University).

GStA PK Geheimes Staatsarchiv Preußischer Kulturbesitz (Secret State Archives Prussian Cultural Heritage Foundation).

KPM Archiv der Königlich Preußischen Porzellanmanufaktur (Archive of the Royal Prussian Porcelain Manufactory).

Publications

Achard, Franz Carl. 1796. *Verzeichnis einer Sammlung Treib-Gewächs-Orangerie-Haus-Pflanzen wie auch im freyen ausdauernder Bäume, Sträucher, perennirender zwey- und einjähriger Gewächse welche in meinem Garten cultivirt und den Liebhabern der Botanik, zum Tausch gegen andere, in diesem Verzeichnisse nicht benannte entweder als Pflanzen oder im Samen angeboten werden.* Berlin: Königliche Ober-Hofbuchdruckerei.

———. 1797. *Kurze für den Landmann überhaupt, besonders aber für die Märkischen Wirthe abgefasste Anleitung zu der Anlage der ergiebigsten künstlichen Wiesen auf Ackerfelder von mittel und leichten Böden.* Berlin: C. L. Sartmann.

Adlung, Alfred. 1928. "Apothekenbesitzer, Apothekengehilfen und -lehrlinge Preußens im Jahre 1798." *Archiv für Sippenforschung und alle verwandten Gebiete*, Heft 5–8: 163–166, 200–203; 229–232, 280–283.

Agricola, Georgius. 1950 (1556). *De re metallica.* Translated from the first Lation edition of 1556 by Herbert Clark Hoover and Lou Henry Hoover, 1912. Reprint, New York: Dover.

Agricola, Georg. 1977 (1557). *Zwölf Bücher vom Berg- und Hüttenwesen.* Translated by Carl Schiffner et al., 1928. Reprint, Munich: dtv.

Alder, Ken. 1997. *Engineering the Revolution: Arms and Enlightenment in France, 1763–1815.* Princeton: Princeton University Press.

Anonymous. 1784. "Vom Herrn M. H. in Berlin." *Chemische Annalen für die Freunde der Naturlehre, Arzneygelahrtheit, Haushaltungskunst und Manufakturen* 1: 342.

Anonymous. 1793a. "Ueber Hrn. Apotheker Merkels in Nürnberg Vorschläge, sich künftig brauchbarere Apotheker-Subjekte zu verschaffen. Von einem der Pharmacie Beflissenen." *Almanach oder Taschenbuch für Scheidekünstler und Apotheker* 1793: 119–132.

Anonymous. 1793b. "Auch ein kleiner Beytrag über den Zustand der Pharmacie in Deutschland." *Almanach oder Taschenbuch für Scheidekünstler und Apotheker* 1793: 49–72.

Anthony, Patrick. 2018. "Mining as the Working World of Alexander von Humboldt's Plant Geography and Vertical Cartography." *Isis* 109: 28–55.

Ash, Eric. 2007. "Amending Nature: Draining the English Fens." In *The Mindful Hand: Inquiry and Invention from the Late Renaissance to Early Industrialization*, edited by Lissa Roberts, Simon Schaffer, and Peter Dear, 117–143. Amsterdam: Koninklijke Nederlandse Akademie van Wetenschappen.

———. 2010a. "Introduction: Expertise and the Early Modern State." In *Expertise: Practical Knowledge and the Early Modern State*, edited by Eric H. Ash. *Osiris* 25: 1–24.

———. ed. 2010b. *Expertise: Practical Knowledge and the Early Modern State. Osiris* 25.

Ashworth, William J. 2017. *The Industrial Revolution: The State, Knowledge and Global Trade.* London: Bloomsbury.

Bacon, Francis. 1986–1994. *The Works of Francis Bacon.* 14 vols. Stuttgart: Frommann-Holzboog. Reprint of the edition by Spedding, Ellis, and Heath (1857–1874).

Banse, Gerhard, and Arnim Grunwald. 2009. "Coherence and Diversity in the Engineering Sciences." In *Philosophy of Technology and Engineering Sciences*, edited by Anthonie Meijers, 155–184. Amsterdam: Elsevier.

Banse, Gerhard, Arnim Grunwald, Wolfgang König, and Günter Ropohl, eds. 2006. *Erkennen und Gestalten: Eine Theorie der Technikwissenschaften.* Berlin: Edition Sigma.

Banse, Gerhard, and Hans-Peter Müller, eds. 2001. *Johann Beckmann und die Folgen: Erfindungen—Versuch der historischen, theoretischen und empirischen Annäherung an einen vielschichtigen Begriff.* Münster: Waxmann.

Barnes, Barry. 2005. "Elusive Memories of Technoscience." *Perspectives on Science* 13: 42–165.

Bartels, Christoph. 1992. *Vom frühneuzeitlichen Montangewerbe zur Bergbauindustrie: Erzbergbau im Oberharz 1635–1866*. Veröffentlichungen aus dem Deutschen Bergbaumuseum No. 54. Bochum: Deutsches Bergbaumuseum Bochum.

———. 2009. "Vermessungswesen, Karten und Pläne im Montanwesen an der Wende zwischen Mittelalter und Neuzeit—Kontinuitätslinien und Entwicklungstendenzen." In *Aufsicht—Ansicht—Einsicht: Neue Perspektiven auf die Kartographie an der Schwelle zur Frühen Neuzeit*, edited by Tanja Michalsky, Felicitas Schmieder, and Gisela Engel, 329–350. Berlin: Trafo Wissenschaftsverlag.

———. 2013. "Der Harzer Oberbergmeister Georg Andreas Stelzner (1725–1825) und die Montanwissenschaften in der zweiten Hälfte des 18. und am Beginn des 19. Jahrhunderts." In *Staat, Bergbau und Bergakademie: Montanexperten im 18. und frühen 19. Jahrhundert*, edited by Hartmut Schleiff and Peter Konečný, 275–288. Stuttgart: Steiner.

Bartels, Christoph, Mariá Ruiz del Árbol, Heleen van Londen, and Almudena Orejas. 2008. *Landmarks—Profiling Europe's Historic Landscapes*. Bochum: Deutsches Bergbau-Museum.

Baumgärtel, Hans. 1963. *Bergbau und Absolutismus: Der Sächsische Bergbau in der zweiten Hälfte des 18. Jahrhunderts und Maßnahmen zu seiner Verbesserung nach dem Siebenjährigen Kriege*. Freiberger Forschungshefte Kultur und Technik D 44. Leipzig: VEB Deutscher Verlag für Grundstoffindustrie.

———. 1965. *Vom Bergbüchlein zur Bergakademie: Zur Entstehung der Bergbauwissenschaften zwischen 1500 und 1750/1770*. Freiberger Forschungshefte D 50. Leipzig: VEB Deutscher Verlag für Grundstoffindustrie.

Baumgärtel, Hans, and Eberhard Wächtler. 1965. "Die Stipendienkasse 1702 bis 1765." In *Bergakademie Freiberg, Festschrift zu ihrer Zweihundertjahrfeier am 13. November 1965*, edited by Rektor und Senat der Bergakademie Freiberg, 60–62. Leipzig: VEB Deutscher Verlag für Grundstoffindustrie.

Bayerl, Günter. 1999. "Der Zugriff auf das Naturreich: Vorindustrielles Gewerbe und Umwelt." In *Johann Beckmann (1739–1811), Beiträge zu Leben, Werk und Wirkung des Begründers der Allgemeinen Technologie*, edited by Günter Bayerl and Jürgen Beckmann, 69–86. Münster: Waxmann.

Bayerl, Günter, and Jürgen Beckmann, eds. 1999. *Johann Beckmann (1739–1811): Beiträge zu Leben, Werk und Wirkung des Begründers der Allgemeinen T1echnologie*. Münster: Waxmann.

Beck, Hanno. 1959. *Alexander von Humboldt: Von der Bildungsreise zur Forschungsreise, 1769–1804*. Wiesbaden: F. Steiner.

———. 1973. *Geographie, Europäische Entwicklung in Texten und Erläuterungen*. Freiburg and Munich: Karl Alber.

Beckmann, Johann. 1777. *Anleitung zur Technologie, oder zur Kentniß der Handwerke, Fabriken und Manufacturen, vornehmlich derer, die mit der Landwirthschaft, Polizey und Cameralwissenschaft in nächster Verbindung stehn.* Göttingen: Vandenhoeck.

Belhoste, Bruno. 2003. *La formation d'une technocratie: l'École polytechnique et ses élèves de la Révolution au Second Empire.* Paris: Belin.

———. 2011. *Paris Savant, parcours et rencontres au temps des Lumières.* Paris: Armand Colin.

———. 2012. "A Parisian Craftsman among the Savants: The Jointer André-Jacob Roubo (1739–1791) and His Works." *Annals of Science* 69 (3): 395–411.

Bennett, Jim. 1987. *The Divided Circle: A History of Instruments for Astronomy, Navigation, and Surveying.* Oxford: Phaidon Press.

Bensaude-Vincent, Bernadette, and Sacha Loeve. 2018. "Toward a Philosophy of Technoscience." In *French Philosophy of Technology: Classical Readings and Contemporary Approaches*, edited by Sacha Loeve, Xavier Guchet, and Bernadette Bensaude-Vincent, 169–186. Cham: Springer.

Beretta, Marco. 2012. "The Rise and Fall of the Glassmaker Paul Bosc d'Antic (1753–1784)." *Annals of Science* 69 (3): 375–393.

———. 2014. "Unveiling Glass's Mysteries: Lavoisier, Loysel and the First Chemical Treatise on Glass (1765–1799)." In *Objects of Chemical Inquiry*, edited by Ursula Klein and Carsten Rheinhardt, 1–20. Sagamore Beach, MA: Watson Publishing International.

Bertucci, Paola. 2017. *Artisanal Enlightenment: Science and the Mechanical Arts in Old Regime France.* New Haven: Yale University Press.

Beuth, Christian Peter Wilhelm. 1822. "Bericht des Geh. Ober-Finanzrathes Beuth an den Herrn Minister für Handel und Gewerbe über die auf dessen Befehl zur Ausbildung der Gewerbetreibenden getroffenen Einrichtungen." *Verhandlungen des Vereins zur Beförderung des Gewerbefleißes in Preußen* 1: 133–142.

Biermann, Kurt-R. 1991. *Beglückende Ermunterung durch die akademische Gemeinschaft: Alexander von Humboldt als Mitglied der Berliner Akademie der Wissenschaften.* Berlin: Akademie Verlag.

Biermann, Kurt-R., Ilse Jahn, and Fritz G. Lange. 1968. *Alexander von Humboldt: Chronologische Übersicht über wichtige Daten seines Lebens.* Berlin: Akademie Verlag.

Blair, Ann. 2006. "Natural Philosophy." In *The Cambridge History of Science*, vol. 3: *Early Modern Science*, edited by Katherine Park and Lorraine Daston, 365–406. Cambridge: Cambridge University Press.

Bleker, Johanna, Petra Lennig, and Thomas Schnalke, eds. 2018. *Tiefe Einblicke: Das Anatomische Theather im Zeitalter der Aufklärung.* Berlin: Kadmos.

Boon, Mieke, and Tarja Knuuttila. 2009. "Models as Epistemic Tools in Engineering Sciences." In *Philosophy of Technology and Engineering Sciences*, edited by Anthonie Meijers, 693–726. Amsterdam: Elsevier.

Boullay, Pierre François Guillaume. 1807. "Mémoire sur le mode de composition des éthers muriatique et acétique." *Annales de Chimie* 63: 90–101.

Braun, Hans-Joachim. 1977. "Methodenprobleme der Ingenieurwissenschaft, 1850–1900." *Technikgeschichte* 44 (1): 1–18.

Bredekamp, Horst, and Adam S. Labuda. 2012. "Kunstgeschichte, Universität, Museum und die Mitte Berlins." In *Geschichte der Universität Unter den Linden*, vol. 1, edited by Heinz-Elmar Tenorth, 237–263. Berlin: Akademie Verlag.

Breil, Hans. 1977. *Friedrich August Alexander Eversmann und die industriell-technologische Entwicklung vornehmlich in Preußen von 1780 bis zum Ausgang der napoleonischen Ära.* Dissertation, Universität Hamburg.

Bret, Patrice. 2002. *L'état, l'armée, la science: l'invention de la recherche publique en France (1763–1830)*. Rennes: Presses Universitaires de Rennes.

Brianta, Donata. 2000. "Education and Training in the Mining Industry, 1750–1860: European Models and the Italian Case." *Annals of Science* 57 (3): 267–300.

Brock, William H. 1993. *The Norton History of Chemistry*. New York: W. W. Norton.

Brose, Eric Dorn. 1993. *The Politics of Technological Change in Prussia: Out of the Shadow of Antiquity, 1809–1848*. Princeton: Princeton University Press.

Brückner, Jutta. 1977. *Staatswissenschaften, Kameralismus und Naturrecht*. Munich: Beck.

Bruhns, Karl. 1969. *Alexander von Humboldt: eine wissenschaftliche Biographie*. 3 vols. Osnabrück (1872): Otto Zeller Verlag.

Buchheim, Gisela, and Rolf Sonnemann. 1990. *Geschichte der Technikwissenschaften*. Leipzig: Edition Leipzig.

Bud, Robert. 2012. "'Applied Science': A Phrase in Search of Meaning." *Isis* 103: 537–545.

———. 2014. "Applied Science in Nineteenth-Century Britain: Public Discourse and the Creation of Meaning, 1817–1876." *History and Technology* 30 (1/2): 3–36.

Buddensieg, Tilmann, Kurt Düwell, and Klaus-Jürgen Sembach, eds. 1987. *Wissenschaften in Berlin, Drei Begleitbände zur Ausstellung "Der Kongress denkt": Disziplinen, Objekte, Gedanken*. Berlin: Gebr. Mann.

Cahan, David. 1989. *An Institute for an Empire: The Physikalisch-Technische Reichsanstalt 1871–1918*. Cambridge: Cambridge University Press.

———. 2018. *Helmholtz: A Life in Science*. Chicago: University of Chicago Press.

Calvör, Henning. 1763. *Acta historico-chronologico-mechanica circa metallurgiam in Hercynia superiori. Oder Historisch-chronologische Nachricht und theoretische und praktische Beschreibung des Maschinenwesens und der Hülfsmittel bey dem Bergbau auf dem Oberharze*. Braunschweig: Verlag der Fürstl. Waisenhaus-Buchhandlung.

Caneva, Kenneth L. 1997. "Physics and Naturphilosophie: A Reconnaissance." *History of Science* 35: 35–106.

Carlowitz, Hannß Carl von. 1713. *Sylvicultura oeconomica, oder, Hausswirthliche Nachricht und naturmässige Anweisung zur wilden Baum-Zucht*. Leipzig: Johann Friedrich Braun.

Chalmers, Alan F. 2017. *One Hundred Years of Pressure: Hydrostatics from Stevin to Newton*. Cham: Springer.

Channell, David F. 1989. *The History of Engineering Science. An Annotated Bibliography*. New York: Garland Publishing.

———. 2009. "The Emergence of the Engineering Sciences: An Historical Analysis." In *Philosophy of Technology and Engineering Sciences*, edited by Anthonie Meijers, 117–154. Amsterdam: Elsevier.

———. 2017. *A History of Technoscience: Erasing Boundaries between Science and Technology*. London: Routledge.

Charpentier, Johann Friedrich Wilhelm von. 1778. *Petrographische Karte des Churfürstenthums Sachsen und der Incorporirten Lande, in welcher durch Farben und Zeichen die Gesteinsarten, durch die an mehrern Orten beygesetzten Zahlen über die nach Barometrischen Beobachtungen gefundenen Höhen dieser Örter über Wittenberg in Pariser Fuss angegeben sind*. N.p.

Clark, William. 2006. *Academic Charisma and the Origins of the Research University*. Chicago: University of Chicago Press.

Clarke, Sabine. 2010. "Pure Science with a Practical Aim: The Meanings of Fundamental Research in Britain, circa 1916–1950." *Isis* 101 (2): 285–311.

Cohen, Floris H. 1994. *The Scientific Revolution: A Historical Inquiry*. Chicago: University of Chicago Press.

Cole, R. J. 1951. "Friedrich Accum (1769–1838): A Biographical Study." *Annals of Science* 7 (2): 128–143.

Condrau, Flurin. 2005. *Die Industrialisierung in Deutschland*. Darmstadt: Wissenschaftliche Buchgesellschaft.

Cook, Harold J. 2007. *Matters of Exchange: Commerce, Medicine, and Science in the Dutch Golden Age*. New Haven: Yale University Press.

Cormack, Lesley B., Steven A. Walton, and John A. Schuster, eds. 2017. *Mathematical Practitioners and the Transformation of Natural Knowledge in Early Modern Europe*. Cham: Springer.

Crell, Lorenz. 1778. "Vorrede." *Chemisches Journal* 1778: 9–20.

Cunningham, Andrew, and Perry Williams. 1993. "De-centring the 'Big Picture': The Origins of Modern Science and the Modern Origins of Science." *British Journal for the History of Science* 26: 407–423.

Damerow, Peter, and Jürgen Renn. 2010. "The Transformation of Ancient Mechanics into a Mechanistic Worldview." In *Transformationen antiker Wissenschaften*, edited by Georg Töpfer and Hartmut Böhme, 239–263. Berlin: de Gruyter.

Dann, Georg Edmund. 1958. *Martin Heinrich Klaproth (1743–1817): Ein deutscher Apotheker und Chemiker, sein Weg und seine Leistung*. Berlin: Akademie-Verlag.

Darmstaedter, Ernst. 1926. *Berg-, Probier- und Kunstbüchlein*. Munich: B. Heller.

Daston, Lorraine. 2017. "The History of Science and the History of Knowledge." *KNOW* 1(1): 131–154.

Daston, Lorraine, and Peter Galison. 2007. *Objectivity*. New York: Zone Books.

Davids, Karel. 2012. "Gatekeeping: Who Defined 'Useful Knowledge' in Early Modern Times?" *History of Technology* 31: 69–88.

Dear, Peter. 1995. *Discipline and Experience: The Mathematical Way in the Scientific Revolution*. Chicago: University of Chicago Press.

———. 2007. "Towards a Genealogy of Modern Science." In *The Mindful Hand: Inquiry and Invention from the Late Renaissance to Early Industrialization*, edited by Lissa Roberts, Simon Schaffer, and Peter Dear, 431–452. Amsterdam: Koninklijke Nederlandse Akademie van Wetenschappen.

Delius, Christoph Traugott. 1773. *Anleitung zu der Bergbaukunst nach ihrer Theorie und Ausübung, nebst einer Abhandlung von den Grundsätzen der Berg-Kammeralwissenschaft*. Vienna: v. Trattern.

Dittrich, Erhard. 1974. *Die deutschen und österreichischen Kameralisten*. Darmstadt: Wissenschaftliche Buchgesellschaft.

Dobbert, Eduard. 1899. "Bauakademie, Gewerbeakademie und Technische Hochschule bis 1884." In *Chronik der Königlichen Technischen Hochschule zu Berlin, 1799–1899*, edited by Rector und Senat der KTH, 11–114. Berlin: Verlag Wilhelm Ernst & Sohn.

Duffy, Christopher. 1974. *The Army of Fredrick the Great*. Vancouver: David & Charles.

Duncan, Alistair M. 1996. *Laws and Order in Eighteenth-Century Chemistry*. Oxford: Clarendon Press.

Dupré, Sven 2017. "Doing It Wrong: The Translation of Artisanal Knowledge and the Codification of Error." In *The Structures of Practical Knowledge*, edited by Matteo Valleriani, 167–188. Cham: Springer.

Dupré, Sven, and Christoph Lüthy, eds. 2011. *Silent Messengers: The Circulation of Material Objects of Knowledge in the Early Modern Low Countries*. New Brunswick, NJ: Transaction Publishers.

Dym, Warren Alexander. 2011. *Divining Science: Treasure Hunting and Earth Science in Early Modern Germany*. Leiden: Brill.

Eckert, Michael. 2002. "Euler and the Fountains of Sanssouci." *Archive of the History of the Exact Sciences* 56: 451–468.

Eichler, Helga. 1974. *Die Preußische Akademie der Wissenschaften zwischen 1740 und 1812; unter besonderer Berücksichtigung ihrer Bedeutung für die Entwicklung der Produktivkräfte*. Dissertation, Humboldt Universität zu Berlin.

Engel, Michael. 2013. "Der Berg- und Hüttenmännische Unterricht in Berlin 1770 bis 1810, die sogenannte Bergakademie." In *Staat, Bergbau und Bergakademie: Montanexperten im 18. und frühen 19. Jahrhundert*, edited by Hartmut Schleiff and Peter Konečný, 95–124. Stuttgart: Steiner.

Engelhardt, Dietrich von. 1990. "Historical Consciousness in the German Romantic Naturforschung." In *Romanticism and the Sciences*, edited by Andrew Cunningham and Nicholas Jardine, 55–68. Cambridge: Cambridge University Press.

Epple, Moritz. 2008. "The Gap between Theory and Practice: Hydrodynamical and Hydraulical Utopias in the 18th Century." In *Philosophies of Technology: Francis Bacon and His Contemporaries*, edited by Claus Zittel, Gisela Engel, Romano Nanni, and Nicole C. Karafyllis, 457–493. Leiden: Brill.

Eytelwein, Johann Albert. 1801. *Handbuch der Mechanik fester Körper und der Hydraulik. Mit vorzüglicher Rücksicht auf ihre Anwendung in der Architektur*. Berlin: Lagarde.

Fechner, Hermann. 1900–1901. "Geschichte des Schlesischen Berg- und Hüttenwesens in der Zeit Friedrich des Grossen, Friedrich Wilhelm's II. und Friedrich Wilhelm's III. 1741 bis 1806." *Zeitschrift für das Berg-, Hütten- und Salinen-Wesen im Preussischen Staate*, vol. 48: 279–401, 310–311 (part 1); vol. 49: 1–86, 243–288, 383–446, 487–569 (part 2).

Felten, Sebastian. 2018. "The History of Science and the History of Bureaucratic Knowledge: Saxon Mining ca. 1770." *History of Science*, online publication DOI: 10.1177/0073275318792451, 1–29.

Fessner, Michael. 2005. "Montanwissenschaften." In *Enzyklopädie der Neuzeit*, vol. 8, edited by Friedrich Jaeger, 768–774. Stuttgart and Weimar: Metzler.

Fischer, F. G. 1820. "Denkschrift auf Klaproth." *Abhandlungen der Königlichen Akademie der Wissenschaften in Berlin*, vol. 1818/19: 11–26.

Fleischer, Alette. 2007. "The Beemster Polder: Conservative Invention and Holland's Great Pleasure Garden." In *The Mindful Hand: Inquiry and Invention from the Late Renaissance to Early Industrialization*, edited by Lissa Roberts, Simon Schaffer, and Peter Dear, 145–166. Amsterdam: Koninklijke Nederlandse Akademie van Wetenschappen.

Forman, Paul. 2007. "The Primacy of Science in Modernity, of Technology in Postmodernity, and of Ideology in the History of Technology." *History and Technology* 23 (1–2): 1–152.

Fourcroy, Antoine-François de, and Louis N. Vauquelin. 1797. "De l'action de l'Acide sulfurique sur l'Alcool, et de la formation de l'Ether." *Annales de Chimie* 23: 203–215.

Frercks, Jan. 2008. "Techniken der Vermittlung. Chemie als Verbindung von Arbeit, Lehre und Forschung am Beispiel von J. F. A. Göttling." *NTM* 16: 279–308.

Freudenthal, Gideon, and Peter McLaughlin. 2009. *The Social and Economic Roots of the Scientific Revolution. Texts by Boris Hessen and Henryk Grossmann*. Dordrecht: Springer.

Freyberg, Bruno von. 1955. *Johann Gottlob Lehmann (1719–1767), Ein Arzt, Chemiker, Metallurg, Bergmann, Mineraloge und grundlegender Geologe*. Erlangen: Universitätsverbund Erlangen.

Freyer, Günter. 1985. "Zur Entstehungsgeschichte geologischer Übersichtskarten von Sachsen, herausgegeben von der 'Sächsischen Geologischen Landesuntersuchung'." In *Leben und Wirken Deutscher Geologen im 18. und 19. Jahrhundert*, edited by Hans Prescher, 357–372. Leipzig: Deutscher Verlag für Grundstoffindustrie.

Fritscher, Bernhard. 2013. "Erdgeschichtsschreibung als montanistische Praxis: Zum nationalen Stil einer 'preußischen Geognosie'." In *Staat, Bergbau und Bergakademie: Montanexperten im 18. und frühen 19. Jahrhundert*, edited by Hartmut Schleiff and Peter Konečný, 205–229. Stuttgart: Steiner.

Frobenius, Siegmund August. 1729–1730. "An Account of a Spiritus Vini Aetherius, Together with Several Experiments Tried Therewith." *Philosophical Transactions* (1683–1775) 36: 283–289.

Füssel, Marian. 2012. "Die Experten, die Verkehrten? Gelehrtensatire als Expertenkritik in der Frühen Neuzeit." In *Wissen maßgeschneidert: Experten und Expertenkulturen im Europa der Vormoderne*, edited by Björn Reich, Frank Rexroth, and Matthias Roick, 269–288. Munich: Oldenburg Verlag.

Gandert, Klaus-Dietrich. 2004. *Vom Prinzenpalais zur Humboldt-Universität. Die historische Entwicklung des Universitätsgebäudes in Berlin mit seinen Gartenanlagen und Denkmälern*. Berlin: Henschel Verlag.

Gaukroger, Stephen. 2006. *The Emergence of a Scientific Culture: Science and the Shaping of Modernity, 1210–1685*. Oxford: Oxford University Press.

———. 2010. *The Collapse of Mechanism and the Rise of Sensibility: Science and the Shaping of Modernity, 1680–1760*. Oxford: Oxford University Press.

———. 2016. *The Natural and the Human: Science and the Shaping of Modernity, 1739–1841*. Oxford: Oxford University Press.

Geheimes Staatsarchiv Preußischer Kulturbesitz, ed. 2014. *Klosterstrasse 36, Sammeln, Ausstellen, Patentieren: Zu den Anfängen Preußens als Industriestaat. Ausstellungskatalog des Geheimen Staatsarchivs Preußischer Kulturbesitz in Zusammenarbeit mit der Kunstbibliothek der Staatlichen Museen zu Berlin*. Berlin: Geheimes Staatsarchiv Preußischer Kulturbesitz.

Gerhard, Carl Abraham. 1771. "Observations physiques et minéralogiques sur les montagnes de la Silésie." *Nouveaux Mémoires de L'Académie Royale des Sciences et Belles Lettres* 1771: 100–111, 112–122.

———. 1773–1776. *Beiträge zur Chymie und Geschichte des Mineralreichs*. 2 vols. Berlin: C. F. Himburg.

———. 1777–1785. *Gabriel Jars Metallurgische Reisen zur Untersuchung und Beobachtung der vornehmsten Berg- und Hüttenwerke in Schweden, Norwegen, Ungarn, Deutschland, Engelland und Schottland, vom Jahr 1757–1769, aus dem Französischen übersetzt von Carl Abraham Gerhard*. 3 vols. Berlin: C. F. Himburg.

———. 1779. "Nouvelle Méthode d'extraire le Bleu royal de toutes sortes de Cobalt à l'usage des Fabriques de Porcelaine." *Nouveaux Mémoires de L'Académie Royale des Sciences et Belles Lettres* 1779: 12–19.

———. 1780. "Considérations générales sur les différences du fer et sur leurs causes." *Nouveaux Mémoires de L'Académie Royale des Sciences et Belles Lettres* 1780: 68–80.

———. 1781a. "Mémoire sur le rapport qu'il y a entre les Terres & les Pierres exposées au feu du fusion, dans des creusets de matières différentes." *Nouveaux Mémoires de L'Académie Royale des Sciences et Belles Lettres* 1781: 80–102.

———. 1781b. "Sur une nouvelle fabrication du verre." *Nouveaux Mémoires de L'Académie Royale des Sciences et Belles Lettres* 1781: 113–124.

———. 1781–1782. *Versuch einer Geschichte des Mineralreichs*. 2 vols. Berlin: C. F. Himburg.

———. 1786. *Grundriß des Mineralsystems zu Vorlesungen*. Berlin: C. F. Himburg.

———. 1797. *Grundriß eines neuen Mineralsystems*. Berlin: Vieweg.

Gillispie, Charles C. 1957a. "The Natural History of Industry." *Isis* 48: 398–407.

———. 1957b. "The Discovery of the Leblanc Process." *Isis* 48: 152–170.

———. 1980. *Science and Polity in France: The End of the Old Regime*. Princeton: Princeton University Press.

———, ed. 1981a. *Dictionary of Scientific Biography*. 16 vols. New York: Charles Scribner's Sons.

———. 1981b. "Bélidor, Bernard Forest de." In *Dictionary of Scientific Biography*, edited by Charles Gillispie, vol. 2, 581–582. New York: Charles Scribner's Sons.

Gilly, David. 1797–1798. *Handbuch der Land-Bau-Kunst vorzüglich in Rücksicht auf die Construction der Wohn- und Wirtschaftsgebäude für angehende Cameral-Baumeister und Oeconomen*. 2 vols. Berlin. Goerge Decker.

Gleditsch, Johann Gottlieb. 1774–1775. *Systematische Einleitung in die neuere, aus ihren eigenthümlichen physikalisch-ökonomischen Gründen hergeleitete Forstwirtschaft*. 2 vols. Berlin: Arnold Wever.

Gmelin, Johann Friedrich. 1786. *Grundsätze der technischen Chemie*. Halle: J. J. Gebauer.

Gohau, Gabriel. 1990. *A History of Geology*. New Brunswick, NJ: Rutgers University Press.

Goldschmidt, Friedrich, and Paul Goldschmidt. 1881. *Das Leben des Staatsrath Kunth*. Berlin: Julius Springer.

Gooday, Graeme. 2012. "'Vague and Artificial': The Historically Elusive Distinction between Pure and Applied Science." *Isis* 103: 546–554.

Göttling, Johann Friedrich A. 1778. *Einleitung in die pharmaceutische Chymie für Lernende*. Altenburg: Richter.

———. 1779. "Chymische Versuche mit der Holzsäure, in Absicht vermittelst derselben eine Naphtha zu verfertigen." *Chemisches Journal für die Freunde der Naturlehre, Arzneygelahrtheit, Haushaltungskunst und Manufacturen* 2: 39–61.

———. 1781a. "Sonderbare Bemerkungen über die Eßig-Naphte." *Almanach oder Taschenbuch für Scheidekünstler und Apotheker* 1781: 4–10.

———. 1781b. "Besondere Bemerkungen bey Verfertigung der Salpeter Naphte." *Almanach oder Taschenbuch für Scheidekünstler und Apotheker* 1781: 39–42.

———. 1783a. "Vorbericht." *Almanach oder Taschenbuch für Scheidekünstler und Apotheker* 1783: n.p.

———. 1783b. "Ameisenaether." *Almanach oder Taschenbuch für Scheidekünstler und Apotheker* 1783: 77–78.

———. 1786. "Vorbericht." *Almanach oder Taschenbuch für Scheidekünstler und Apotheker* 1786: n.p.

———. 1787. "Vorbericht." *Almanach oder Taschenbuch für Scheidekünstler und Apotheker* 1787: n.p.

———. 1792. "Vorbericht." *Almanach oder Taschenbuch für Scheidekünstler und Apotheker* 1792: n.p.

Grashof, Franz. 1864. "Ueber die Organisation von polytechnischen Schulen zu Grunde zu legenden Prinzipien." *Zeitschrift des Vereins Deutscher Ingenieure* 8: 591–616.

———. 1876. "Über die wünschenswerthe Entwicklung der deutschen technischen Hochschulen und über Staatseinrichtungen zu geeigneter Verwendung akademisch gebildeter Techniker im öffentlichen Dienste." *Zeitschrift des Vereins Deutscher Ingenieure* 20, 624–640.

Grimm, Jacob, and Wilhelm Grimm. 1854–1956. *Deutsches Wörterbuch*. Leipzig: S. Hirzel.

Guntau, Martin. 1984. *Abraham Gottlob Werner*. Leipzig: Teubner.

Hahn, Hans-Werner. 2011 (1998). *Die Industrielle Revolution in Deutschland*. Munich: Oldenburg Verlag.

Hahn, Roger. 1971. *The Anatomy of a Scientific Institution: The Paris Academy of Sciences, 1666–1803*. Berkeley: University of California Press.

Hall, Rupert A. 1959. "The Scholar and the Craftsman in the Scientific Revolution." In *Critical Problems in the History of Science*, edited by Marshall Clagett, 3–23. Reprint, Madison: University of Wisconsin Press, 1969.

———. 1963. "Merton Revisited or Science and Society in the Seventeenth Century." *History of Science* 2: 1–16.

Hannesen, Hans Gerhard. 2005. *Die Akademie der Künste in Berlin: Facetten einer 300jährigen Geschichte*. Berlin: Akademie der Künste.

Hansson, Sven Ove. 2007. "What Is Technological Science?" *Studies in History and Philosophy of Science* 38: 523–527.

Harkness, Deborah E. 2007. *The Jewel House: Elizabethan London and the Scientific Revolution*. New Haven: Yale University Press.

Harnack, Adolf. 1900. *Geschichte der Königlich-Preussischen Akademie der Wissenschaften zu Berlin*. 4 vols. Berlin: Reichsdruckerei.

Harwood, Jonathan. 2005. *Technology's Dilemma: Agricultural Colleges between Science and Practice in Germany, 1860–1934*. Frankfurt: Peter Lang.

———. 2006. "Engineering Education between Science and Practice: Rethinking the Historiography." *History and Technology* 22 (1): 53–79.

———. 2010. "Understanding Academic Drift: On the Institutional Dynamics of Higher Technical and Professional Education." *Minerva* 48: 413–427.

Hasse, T. L. 1848. *Denkschrift zur Erinnerung an die Verdienste des in Dresden am 30. Juni 1817 verstorbenen K. G. Bergrath's Werner und an die Fortschritte bei der Bergakademie zu Freiberg*. Dresden and Leipzig: Arnoldische Buchhandlung.

Heinitz, Friedrich Anton von. 1963 (1771). "Bericht der Revisionskommission vom 2.3. 1771." In Hans Baumgärtel, *Bergbau und Absolutismus: Der Sächsische Bergbau in der zweiten Hälfte des 18. Jahrhunderts und Maßnahmen zu seiner Verbesserung nach dem Siebenjährigen Kriege.* Freiberger Forschungshefte Kultur und Technik D, Heft 44, 121–190. Leipzig: VEB Deutscher Verlag für Grundstoffindustrie.

Henderson, William O. 1958. *The State and the Industrial Revolution in Prussia, 1740– 1870.* Liverpool: Liverpool University Press.

———. 1961. "Die Struktur der preußischen Wirtschaft um 1786." *Zeitschrift für die Gesamte Staatswirtschaft* 117: 292–319.

Henning, Friedrich Wilhelm. 1973. *Die Industrialisierung in Deutschland 1800 bis 1914.* Paderborn: Schöningh.

Henschel, Klaus. 2011. "Von der Werkstoffforschung zur Materials Science." *NTM* 19: 5–40.

Hermbstaedt, Sigismund Friedrich. 1790. "Nachricht von einer chemischen Pensionsanstalt für Jünglinge, die sich zu praktischen Chemikern bilden wollen." *Chemische Annalen* 1790, pt. 1: 94–96.

Herrmann, Walther. 1953. *Bergbau und Kultur: Beiträge zur Geschichte des Freiberger Bergbaus und der Bergakademie.* Freiberger Forschungshefte Kultur und Technik D, Heft 2. Berlin: Akademie Verlag.

———. 1962. *Bergrat Henckel, ein Wegbereiter der Bergakademie.* Freiberger Forschungshefte Kultur und Technik D, Heft 37. Berlin: Akademie Verlag.

Herzfeld, Erika. 1994. *Preußische Manufakturen: Großgewerbliche Fertigung von Porzellan, Seide, Gobelins, Uhren, Tapeten, Waffen u.a. im 17. und 18. Jahrhundert in und um Berlin.* Berlin: Verlag der Nation.

Hilaire Pérez, Liliane. 2000. *L'invention technique au siècle des Lumières.* Paris: Albin Michel.

———. 2008. "Inventing a World of Guilds: Silk Fabrics in Eighteenth-Century Lyon." In *Guilds, Innovation, and the European Economy, 1400–1800,* edited by S. R. Epstein and Maarten Prak, 232–263. Cambridge: Cambridge University Press.

Hilaire-Pérez, Liliane, and Catherine Verna. 2006. "Dissemination of Technical Knowledge in the Middle Ages and the Early Modern Era: New Approaches and Methodological Issues." *Technology and Culture* 47: 536–565.

Hofbauer, Gottfried. 2015. *Die geologische Revolution: Wie die Entdeckung der Erdgeschichte unser Denken veränderte.* Darmstadt: Wissenschaftliche Buchgesellschaft.

Homburg, Ernst. 1999. "The Rise of Analytical Chemistry and Its Consequences for the Development of the German Chemical Profession (1780–1860)." *Ambix* 46 (1): 1–32.

Hoppe, Günter. 1989. "Martin Heinrich Klaproth (1743–1817) als Mineralchemiker und Mineralsammler." *Der Aufschluß* 40: 201–214.

Hörmann, Johannes. 1898. "Die Königliche Hofapotheke in Berlin (1598–1898)." *Hohenzollern-Jahrbuch* 1898: 208–226.

Hottois, Gilbert. 2018. "Technoscience: From the Origin of the Word to Its Current Uses." In *French Philosophy of Technology: Classical Readings and Contemporary Approaches*, edited by Sacha Loeve, Xavier Guchet, and Bernadette Bensaude-Vincent, 121–138. Cham: Springer.

Houkes, Wybo. 2009. "The Nature of Technological Knowledge." In *Philosophy of Technology and Engineering Sciences*, edited by Anthonie Meijers, 309–350. Amsterdam: Elsevier.

Hufbauer, Karl. 1982. *The Formation of the German Chemical Community (1720–1795)*. Berkeley: University of California Press.

Hughes, Sally Smith. 2001. "Making Dollars out of DNA: The First Major Patent in Biotechnology and the Commercialization of Molecular Biology, 1974–1980." *Isis* 92: 541–575.

Humboldt, Alexander von. 1792. "Versuch über einige physikalische und chemische Grundsätze der Salzwerkskunde." *Bergmännisches Journal* 5: 1–45, 97–141.

———. 1795. "Ueber die Grubenluft und die Verbreitung des Kohlenstoffs in geognostischer Sicht (aus einem Briefe an Hrn. Prof. Lampadius)." *Chemische Annalen für die Freunde der Naturlehre, Arzneygelahrtheit, Haushaltungskunst, und Manufakturen* 2: 99–119.

———. 1796. "Ueber die einfache Vorrichtung durch welche sich Menschen stundenlang in irrespirablen Gasarten, ohne Nachtheil der Gesundheit, und mit brennenden Lichtern aufhalten können, oder vorläufige Anzeige einer Rettungsflasche und eines Lichthalters, aus einem Briefe des Hrn. Oberbergraths von Humboldt an den Herrn Berghauptmann von Trebra." *Chemische Annalen für die Freunde der Naturlehre, Arzneygelahrtheit, Haushaltungskunst, und Manufakturen* 2: 99–110, 195–210.

———. 1797. "Kurze Darstellung der gegenwärtigen Verhälntiße des Bergbaus in den Fränkischen Fürstentümern." GStA PK, II. HA Gen. Dir. Ansbach-Bayreuth VII, Nr. 34a, 2–30.

———. 1799. *Ueber die unterirdischen Gasarten und die Mittel ihren Nachtheil zu vermindern: Ein Beytrag zur Physik der praktischen Bergbaukunde*. Braunschweig: Vieweg.

———. 1959 (1792). *Bericht über den Zustand des Bergbaus und Hütten-Wesens in den Fürstentümern Bayreuth und Ansbach in Jahre 1792*. Edited with an introduction by Herbert Kühnert and Oskar Oelsner. Berlin: Akademie Verlag.

———. 1973. *Die Jugendbriefe Alexander von Humboldts 1787–1799*. Edited and annotated by Ilse Jahn and Fritz G. Lange. Berlin: Akademie Verlag.

Ihde, Don, and Evan Selinger, eds. 2003. *Chasing Technoscience: Matrix for Materiality*. Bloomington: Indiana University Press.

Inkster, Ian. 1991. *Science and Technology in History: An Approach to Industrial Development*. New Brundwick, NJ: Rutgers University Press.

———. 2006. "Potentially Global: 'Useful and Reliable Knowledge' and Material Progress in Europe, 1474–1914." *International History Review* 28: 237–286.

———. 2012. "Introduction: 'Useful Knowledge' Reconsidered." In *Conceptualizing the Production and Diffusion of Useful and Reliable Knowledge in Early Modern Europe*, edited by Ian Inkster, special issue of *History of Technology* 31: 1–4.

Jackson, Catherine M. 2014. "The Curious Case of Coniine: Constructive Synthesis and Aromatic Structure Theory." In *Objects of Chemical Inquiry*, edited by Ursula Klein and Carsten Rheinhardt, 61–101. Sagamore Beach, MA: Watson Publishing International.

Jacob, Margaret C. 1997. *Scientific Culture and the Making of the Industrial West*. New York: Oxford University Press.

———. 2014. *The First Knowledge Economy. Human Capital and the European Economy, 1750–1850*. Cambridge: Cambridge University Press.

Jacob, Margaret C., and Larry Stewart. 2004. *Practical Matter: Newton's Science in the Service of Industry and Empire, 1687–1851*. Cambridge, MA: Harvard University Press.

Jaeckel, Ralph. 2000. "'Bauen wie auf der Friedrichstadt': Das Retablissement der brandenburg-preußischen Provinzialstädte im 18. Jahrhundert." In *Mathematisches Calcul und Sinn für Ästhetik: Die preußische Bauverwaltung 1770–1848, Ausstellung des Geheimen Staatsarchivs Preußischer Kulturbesitz in Zusammenarbeit mit der Kunstbibliothek der Staatlichen Museen zu Berlin Preußischer Kulturbesitz*, edited by Reinhart Strecke, Christof Baier, Elke Blauert, et al., 11–24. Berlin: Duncker & Humblot.

Johnson, Hubert C. 1975. *Frederick the Great and His Officials*. New Haven: Yale University Press.

Jones, Peter M. 2008. *Industrial Enlightenment: Science, Technology and Culture in Birmingham and the West Midlands, 1760–1820*. Manchester: Manchester University Press.

———. 2016. *Agricultural Enlightenment: Knowledge, Technology, and Nature, 1750–1840*. Oxford: Oxford University Press.

Kahlow, Andreas. 2000. "Die ersten Jahre der Berliner Bauakademie, Vorgeschichte und Zeitbild um 1800." In *1799–1999: Von der Bauakademie zur Technischen Universität Berlin, Geschichte und Zukunft*, edited by Karl Schwarz, 32–55. Ulm: Ernst und Sohn.

Kieffer, Fanny. 2014. "The Laboratories of Art and Alchemy at the Uffizi Gallery in Renaissance Florence: Some Material Aspects." In *Laboratories of Art: Alchemy and Art Technology from Antiquity to the 18th Century*, edited by Sven Dupré, 105–127. Heidelberg: Springer.

Klaproth, Martin Heinrich. 1788–1789. "Über die Anwendbarkeit der Platina zu Verzierungen auf Porcelan." *Mémoires de L'Académie Royale des Sciences et Belles-Lettres* 1788–1789: 12–15.

———. 1789a. "Chemische Untersuchung des Uranits, einer neuentdeckten metallischen Substanz." *Annalen der Chemie* 1789, Teil 2: 387–403.

———. 1789b. "Mémoire chimique et minéralogique sur l'Urane." *Mémoires de L'Académie Royale des Sciences et Belles Lettres* 1786/87: 160–174.

———. 1789c. "Kleine Mineralogische Beyträge." *Annalen der Chemie* 1789, Teil 1: 7–12.

———. 1789d. *Vorlesungen über Experimental-Chemie, nach einer Abschrift aus dem Jahre 1789*, edited by Rüdiger Stolz, Peter Lange, and Rita Schwertner. Berlin: Verlag für Wissenschafts- und Regionalgeschichte M. Engel, 1993.

Klein, Ursula. 1994. *Verbindung und Affinität: Die Grundlegung der neuzeitlichen Chemie an der Wende vom 17. zum 18. Jahrhundert*. Basel: Birkhäuser.

———. 1995. "E. F. Geoffroy's Table of Different 'Rapports' Observed Between Different Chemical Substances—A Reinterpretation." *Ambix* 42 (2): 79–100.

———. 1996. "Experiment, Spiritus und okkulte Qualitäten in der Philosophie Francis Bacons." *Philosophia Naturalis* 33 (2): 289–314.

———. 2005a. "Technoscience avant la lettre." *Perspectives on Science* 13 (2): 226–266.

———. 2005b. "Introduction: Technoscientific Productivity." *Perspectives on Science* 13 (2): 139–141.

———. 2007a. "Apothecary Shops, Laboratories and Chemical Manufacture in Eighteenth-Century Germany." In *The Mindful Hand: Inquiry and Invention from the Late Renaissance to Early Industrialisation*, edited by Lissa Roberts, Simon Schaffer, and Peter Dear, 246–276. Amsterdam: Koninklijke Nederlandse Akademie van Wetenschappen.

———. 2007b. "Apothecary-Chemists in Eighteenth-Century Germany." In *New Narratives in Eighteenth Century Chemistry*, edited by Lawrence M. Principe, 97–137. Dordrecht: Springer.

———. 2008a. "Die technowissenschaftlichen Laboratorien der Frühen Neuzeit." *NTM* 16: 5–33.

———. 2008b. "The Laboratory Challenge: Some Revisions of the Standard Picture of Early Modern Experimentation." *Isis* 99 (4): 769–782.

———. 2010a. "Ein Bergrat, zwei Minister und sechs Lehrende, Versuche zur Gründung einer Bergakademie in Berlin um 1770." *NTM* 18 (4): 437–468.

———. 2010b. "Chemische Wissenschaft und Technologie in der Gründungsphase der Berliner Universität." In *Geschichte der Universität zu Berlin, 1810–2010*, edited by Rüdiger von Bruch and Elmar Tenorth, vol. 4, 447–464. Berlin: Akademie Verlag.

———. 2010c. "Blending Technical Innovation and Learned Natural Knowledge: The Making of Ethers." In *Materials and Expertise in Early Modern Europe: Between Market and Laboratory*, edited by Ursula Klein and Emma C. Spary, 125–157. Chicago: University of Chicago Press.

———. 2012a. "The Prussian Mining Official Alexander von Humboldt." *Annals of Science* 69 (2): 27–68.

———, ed. 2012b. "Artisanal-Scientific Experts in Eighteenth-Century France and Germany." Special Issue of *Annals of Science* 69 (3).

———. 2012c. "Savant Officials in the Prussian Mining Administration." *Annals of Science* 69 (3): 349–374.

———. 2013. "Chemical Experts at the Royal Prussian Porcelain Manufactory." *Ambix* 60: 99–121.

———. 2014a. "Klaproth's Discovery of Uranium." In *Objects of Chemical Inquiry*, edited by Ursula Klein and Carsten Rheinhardt, 21–46. Sagamore Beach, MA: Watson Publishing International.

———. 2014b. "Depersonalizing the Arcanum." *Technology and Culture* 55 (3): 591–621.

———. 2014c. "Chemical Expertise: Chemistry at the Royal Prussian Porcelain Manufactory." *Osiris* 29: 262–282.

———. 2015. *Humboldts Preußen: Wissenschaft und Technik im Aufbruch*. Darmstadt: Wissenschaftliche Buchgesellschaft.

———. 2016a. *Nützliches Wissen: Die Erfindung der Technikwissenschaften*. Göttingen: Wallstein.

———. 2016b. "Kuhn in the Cold War." In *Shifting Paradigms: Thomas S. Kuhn and the History of Science*, edited by Alexander Blum, Kostas Gavroglu, Christian Joas, and Jürgen Renn, 115–121. Berlin: Edition Open Access.

———. 2017. "Hybrid Experts." In *The Structures of Practical Knowledge*, edited by Matteo Valleriani, 287–306. Cham: Springer.

Klein, Ursula, and Wolfgang Lefèvre. 2007. *Materials in Eighteenth-Century Science: A Historical Ontology*. Cambridge, MA: MIT Press.

Klein, Ursula, and Emma C. Spary, eds. 2010a. *Materials and Expertise in Early Modern Europe: Between Market and Laboratory*. Chicago: University of Chicago Press.

Klein, Ursula, and Emma C. Spary. 2010b. "Introduction: Why Materials." In *Materials and Expertise in Early Modern Europe: Between Market and Laboratory*, edited by Ursula Klein and Emma C. Spary, 1–28. Chicago: University of Chicago Press.

Kleinert, Andreas. 2013. "Johann Joachim Lange (1699–1765): Ein Hallescher Mathematikprofessor als Pionier der Montanwissenschaften." In *Staat, Bergbau und*

Bergakademie: Montanexperten im 18. und frühen 19. Jahrhundert, edited by Hartmut Schleiff and Peter Konečný, 194–204. Stuttgart: Steiner.

Klemm, Friedrich. 1966. "Ernst Gottfried Fischer: ein mitteldeutscher Mathematiker, Physiker und Schulmann." *Die Mitte* 2: 27–74.

Klemm, Volker, and Günther Meyer. 1968. *Albrecht Daniel Thaer, Pionier der Landwirtschaftswissenschaften in Deutschland.* Halle: Niemeyer.

Klengel, Bernd. 2010. *Über Galvanismus und deutsche Träumereien: Zur Rezeption romantischer Naturforschung in Frankreich zwischen 1800 und 1820.* Stuttgart: Steiner.

Klosa, Achim M. 2009. *Johann Christian Wiebleb (1732–1800): Eine Ergobiographie der Aufkläring.* Stuttgart: Wissenschaftliche Verlagsgesellschaft.

Kocka, Jürgen. 1969. *Unternehmensverwaltung und Angestelltengesellschaft am Beispiel Siemens 1847–1914.* Stuttgart: Klett.

Köhl, Oskar 1913. *Zur Geschichte des Bergbaus im vormaligen Fürstentume Kulmbach-Bayreuth mit besonderer Berücksichtigung der zum Frankenwalde gehörigen Gebiete, eine kulturgeschichtliche Studie.* Hof: Kommissions-Verlag der Wilhelm Kleinschmidt'schen Buchhandlung.

Kohler, Robert E., and Kathryn M. Olesko, eds. 2012. "Clio Meets Science: The Challenges of History." *Osiris* 27.

Kolbe, Gustav. 1863. *Geschichte der Königlichen Porcellanmanufaktur zu Berlin.* Berlin: Verlag der Königlichen Geheimen Ober-Hofbuchdruckerei.

Köllmann, Erich, and Margarete Jarchow. 1987. *Berliner Porzellan, Textband.* Munich: Klinkhardt und Biermann.

Konečný, Peter. 2013. "Die montanistische Ausbildung in der Habsburger Monarchie, 1763–1848." In *Staat, Bergbau und Bergakademie: Montanexperten im 18. und frühen 19. Jahrhundert,* edited by Hartmut Schleiff and Peter Konečný, 95–124. Stuttgart: Steiner.

König, Wolfgang. 1981. "Stand und Aufgaben der Forschung zur Geschichte der deutschen Polytechnischen Schulen und Technischen Hochschulen im 19. Jahrhundert." *Technikgeschichte* 48 (1): 47–67.

———. 1995. *Technikwissenschaften: Die Entstehung der Elektrotechnik aus Industrie und Wissenschaft zwischen 1880 und 1914.* Chur: Fakultas.

———. 1998. "Zwischen Verwaltungsstaat und Industriegesellschaft: Die Gründung höherer Bildungsstätten in Deutschland in den ersten Jahrzehnten des 19. Jahrhunderts." *Berichte zur Wissenschaftsgeschichte* 21: 115–122.

———. 2000. "Die Technikerbewegung und das Promotionsrecht der Technischen Hochschulen." In *1799–1999: Von der Bauakademie zur Technischen Universität Berlin, Geschichte und Zukunft,* edited by Karl Schwarz, 123–129. Berlin: Ernst & Sohn Verlag.

———. 2006a. "Vom Staatsdiener zum Industrieangestellten: Die Ingenieure in Frankreich und in Deutschland, 1750–1945." In *Geschichte des Ingenieurs, ein Beruf in sechs Jahrtausenden*, edited by Walter Kaiser and Wolfgang König, 179–231. Munich: Hanser.

———. 2006b. "Geschichte der Technikwissenschaften." In *Erkennen und Gestalten: eine Theorie der Technikwissenschaften*, edited by Gerhard Banse, Arnim Grunwald, Wolfgang König, and Günter Ropohl, 24–36. Berlin: Edition Sigma.

Kopp, Hermann. 1843–1847. *Geschichte der Chemie*. 4 vols. Reprint, Braunschweig and Hildesheim: Olms, 1966.

Koselleck, Reinhart. 1967. *Preußen zwischen Reform und Revolution: Allgemeines Landrecht, Verwaltung und soziale Bewegung von 1791 bis 1848*. Suttgart: Klett-Cotta.

Krige, John. 2006. *American Hegemony and the Postwar Reconstruction in Europe*. Cambridge, MA: MIT Press.

Kroker, Werner. 1972. "Aspekte der Entwicklung des Markscheidewesens am Oberharz." *Technikgeschichte* 39: 280–301.

Krünitz, Johann Georg. 1773–1858. *Oeconomische Encyclopädie, oder allgemeines System der Land-, Haus- und Staats-Wirthschaft in alphabetischer Ordnung*. 242 parts. Berlin: J. Pauli.

Krusch, Paul. 1904. *Die Geschichte der Bergakademie zu Berlin von ihrer Gründung im Jahre 1770 bis zur Neueinrichtung im Jahre 1860*. Berlin: Königl. Geologische Landesanstalt und Bergakademie.

Laboulais, Isabelle. 2012. La Maison des mines: La genèse révolutionnaire d'un corps d'ingénieurs civils (1794–1814). Rennes: Presses Universitaires de Rennes.

Laitko, Hubert. 2002. "Die Disziplin als Strukturprinzip und Entwicklungsform der Wissenschaft." In *Die Entstehung Biologischer Disziplinen*, edited by Ekkehard Höxtermann and Uwe Hoßfeld, 19–55. Berlin: Verlag für Wissenschaft und Bildung.

Landecker, Hannah. 2007. *Culturing Life: How Cells Became Technologies*. Cambridge, MA: Harvard University Press.

Langins, Janis. 2004. *Conserving the Enlightenment: French Military Engineering from Vauban to the Revolution*. Cambridge, MA: MIT Press.

Latour, Bruno. 1987. *Science in Action: How to Follow Scientists and Engineers Through Society*. Cambridge, MA: Harvard University Press.

Laudan, Rachel. 1987. *From Mineralogy to Geology: The Foundations of a Science, 1650–1830*. Chicago: University of Chicago Press.

Layton, Edwin T., Jr. 1971. "Mirror-Image Twins: The Communities of Science and Technology in 19th-Century America." *Technology and Culture* 12 (4): 562–580.

———. 1974. "Technology as Knowledge." *Technology and Culture* 15 (1): 31–41.

Lefèvre, Wolfgang. 2001. "Galileo Engineer: Art and Modern Science." In *Galileo in Context*, edited by Jürgen Renn, 11–27. Cambridge: Cambridge University Press.

———, ed. 2004. *Picturing Machines 1400–1700*. Cambridge, MA: MIT Press.

———. 2005. "Science as Labor." *Perspectives on Science* 13 (2): 194–225.

Lehman, Christine. 2012. "Pierre-Joseph Macquer: An Eighteenth-Century Artisanal-Scientific Expert." *Annals of Science* 69 (3): 307–333.

Lehmann, Herbert. 1936. *Das Collegium medico-chirurgicum in Berlin als Lehrstätte der Botanik und der Pharmazie*. Berlin: Triltsch und Huther.

Lehmann, Johann Gottlob. 1752. "Ohnmaßgeblicher Vorschlag, auf was Art und Weise man zu einer genauern Entdeckung der unter der Erde verborgenen Dinge, oder kurz zu sagen, zu einer unterirdischen Erdbeschreibung gelangen könne." *Physikalische Belustigungen* 2, 11. St.: 27–42.

———. 1756. *Versuch einer Geschichte von Flötz-Gebürgen betreffend deren Entstehung, Lage, darinn befindliche Metallen, Mineralien und Fossilien, größtentheils aus eigenen Wahrnehmungen, chymischen und physicalischen Versuchen, und aus den Grundsätzen der Natur-Lehre hergeleitet*. Berlin: Klüter.

Lenz, Max. 1910–1918. *Geschichte der Königlichen Friedrich-Wilhelms-Universität zu Berlin*. 5 vols. Halle: Buchhandlung des Waisenhauses.

Lindenfeld, David F. 1997. *The Practical Imagination: The German Sciences of State in the Nineteenth Century*. Chicago: University of Chicago Press.

Long, Pamela O. 2011. *Artisan/Practitioners and the Rise of the New Sciences, 1400–1600*. Corvallis: Oregon State University Press.

Lowengard, Sarah. 2005. *The Creation of Color in Eighteenth-Century Europe*. New York: Columbia University Press.

Lucier, Paul. 2012. "The Origins of Pure and Applied Science in Gilded Age America." *Isis* 103: 527–536.

Lundgreen, Peter. 1975. *Techniker in Preussen während der frühen Industrialisierung: Ausbildung und Berufsfeld einer entstehenden sozialen Gruppe*. Berlin: Colloquium Verlag.

———. 1990. "Engineering Education in Europe and the U.S.A., 1750–1930: The Rise to Dominance of School Culture and the Engineering Professions." *Annals of Science* 47: 33–75.

Lyotard, François. 1984. *The Postmodern Condition: A Report on Knowledge*. Minneapolis: University of Minnesota Press.

Macquer, Pierre Joseph. 1766. *Dictionnaire de Chymie, contenant la Théorie et la Pratique de cette Science, son application à la Physique, à l'Histoire Naturelle, à la Médicine et à l'Economie animale*. 2 vols. Paris: Lacombe.

Matschoß, Conrad. 1911. "Geschichte der Königlich Preußischen Technischen Deputation für Gewerbe, zur Erinnerung an das 100jährige Bestehen, 1811–1911." *Beiträge zur Geschichte der Technik und Industrie* 3: 239–275.

———. 1921. *Preussens Gewerbeförderung und ihre grossen Männer, dargestellt im Rahmen der Geschichte des Vereins zur Beförderung des Gewerbefleisses, 1821–1921.* Berlin: Verlag des Vereins deutscher Ingenieure.

Mauskopf, Seymour. 2010. "The Crisis of English Gundpowder in the Eighteenth Century." In *Materials and Expertise in Early Modern Europe: Between Market and Laboratory*, edited by Ursula Klein and Emma C. Spary, 288–320. Chicago: University of Chicago Press.

Meijers, Anthonie, ed. 2009. *Philosophy of Technology and Engineering Sciences*. Amsterdam: Elsevier.

Merton, Robert K. 1970 (1938). *Science, Technology, and Society in Seventeenth-Century England*. New York: Harper Torchbooks.

Meyer, Torsten. 2012. "The Science of Building as a Polytechnical Discipline in the 19th Century." In *Nuts and Bolts of Construction History: Culture, Technology and Society*, edited by Robert Carvais, André Guillerme, Valérie Nègre, and Joel Sakarovitch, vol. 1, 21–28. Paris: Picard.

Meyer, Torsten, and Marcus Popplow. 2004. "To Employ Each of Nature's Products in the Most Favorable Way Possible: Nature as Commodity in Eighteenth-Century German Economic Discourse." *Historical Social Research* 29 (4): 4–40.

Mieck, Ilja. 1965a. "Sigismund Friedrich Hermbstaedt (1760 bis 1833), Chemiker und Technologe in Berlin." *Technikgeschichte* 32 (4): 325–382.

———. 1965b. *Preussische Gewerbepolitik in Berlin, 1806–1844.* Berlin: Walter de Gruyter.

Miller, David Philip. 2009. *James Watt, Chemist: Understanding the Origins of the Steam Age*. London: Pickering & Chatto.

Milly, Nicolas-Christiern de Thy, Comte de. 1774. *Die Kunst das ächte Porcellan zu verfertigen*. Translated and annotated by Daniel Gottfried Scherber. Königsberg and Leipzig: Johann Jakob Kanter. Reprint, Leipzig: Zentralantiquariat der Deutschen Demokratischen Republik, 1977.

Mokyr, Joel. 1990. *The Lever of the Riches: Technological Creativity and Economic Progress*. New York: Oxford University Press.

———. 2002. *The Gifts of Athena: Historical Origins of the Knowledge Economy*. Princeton: Princeton University Press.

———. 2005. "The Intellectual Origins of Modern Economic Growth." *Journal of Economic History* 65 (2): 285–351.

Morel, Thomas. 2015. "Le microcosme de la géométrie souterraine: échanges et transmissions en mathématique pratique." *Philosophia Scientiae* 19 (2): 5–24.

Müller, Hans. 1896. *Die Königliche Akademie der Künste zu Berlin 1696 bis 1896*. Berlin: Bong.

Müller, Hans-Heinrich. 2002. *Franz Carl Achard (1753–1821), Biographie*. Berlin: A. Bartens.

Müller-Wille, Staffan, and Katrin Böhme. 2012. "Biologie: Wissenschaft vom Werden, Wissenschaft im Werden." In *Geschichte der Universität Unter den Linden*, edited by Heinz-Elmar Tenorth, vol. 1, 425–446. Berlin: Akademie Verlag.

Multhauf, Robert P. 1966. *The Origins of Modern Chemistry*. London: Oldbourne.

Musson, Albert Edward, and Eric Robinson. 1969. *Science and Technolog in the Industrial Revolution*. Manchester: Manchester University Press.

Nieto-Galan, Agustí. 2001. *Colouring Textiles: A History of Natural Dyestuffs in Industrial Europe*. Dordrecht: Kluwer.

Nordmann, Alfred. 2011. "The Age of Technoscience." In *Science Transformed? Debating Claims of an Epochal Break*, edited by Alfred Nordmann, Hans Radder, and Gregor Schiemann, 19–30. Pittsburgh: University of Pittsburgh Press.

Nordmann, Alfred, Hans Radder, and Gregor Schiemann. 2011. "Science after the End of Science? An Introduction to the 'Epochal Break Thesis'." In *Science Transformed? Debating Claims of an Epochal Break*, edited by Alfred Nordmann, Hans Radder, and Gregor Schiemann, 1–15. Pittsburgh: University of Pittsburgh Press.

Nottebohm, Friedrich W. 1871. *Chronik der Königlichen Gewerbe-Akademie zu Berlin*. Berlin: Königliche Geheime Ober-Hofdruckerei.

Nowotny, Helga, Peter Scott, and Michael Gibbons. 2001. *Rethinking Science: Knowledge and the Public in an Age of Uncertainty*. Cambridge, UK: Polity Press.

Nummedal, Tara. 2007. *Alchemy and Authority in the Holy Roman Empire*. Chicago: University of Chicago Press.

Oldroyd, David. 1996. *Thinking about the Earth: A History of Ideas in Geology*. London: Athlone Press.

Olesko, Kathryn M. 2009. "Geopolitics and Prussian Technical Education in the Late Eighteenth Century." *Actes d'Història de la Ciència i de la Tècnica*, n.s. 2 (2): 11–44.

———. 2020. "Germany." In *The Cambridge History of Science*, vol. 8: *Modern Science in National, Transnational, and Global Context*, edited by Hugh Richard Slotten, Ronald L. Numbers, and David N. Livingstone, 233–277. Cambridge: Cambridge University Press.

Oppel, Friedrich Wilhelm von. 1769. *Bericht vom Bergbau. Ursprünglich niedergeschrieben von Johann Gottlieb Kern, churfürstlich sächsischer Edelstein-Inspektor, bearbeitet und als Lehrbuch herausgegeben von Friedrich Wilhelm von Oppel, churfürstlich sächsischer Oberberghauptmann.* Freyberg. Reprint, with an introduction by Ernst-Ulrich Reuther, Essen: Verlag Glückauf GmbH, 1992.

Ospovat, Alexander M. 1971. "Notes on the Text." In *Short Classification and Description of the Various Rocks*, by Abraham Gottlob Werner, translated and with an introduction and notes by Alexander M. Ospovat, 97–144. New York: Hafner.

Osteroth, Dieter. 1985. *Soda, Teer und Schwefelsäure: der Weg zur Großindustrie.* Hamburg: Rowohlt.

Park, Katherine, and Lorraine Daston. 2006. "Introduction: The Age of the New." In *The Cambridge History of Science*, vol. 3: *Early Modern Science*, edited by Katherine Park and Lorraine Daston, 1–20. Cambridge: Cambridge University Press.

Peithner, Johann Thaddäus Anton. 1769–1770. *Erste Gründe der Bergwerkswissenschaften aus denen physisch-metallurgischen Vorlesungen*, 2 vols. Prague: J. J. Clauser.

Phillips, Denise. 2012. *Acolytes of Nature: Defining Natural Science in Germany, 1770–1850.* Chicago: University of Chicago Press.

Pickstone, John V. 2000. *Ways of Knowing: A New History of Science, Technology, and Medicine.* Manchester: Manchester University Press.

Picon, Antoine. 1992. *L'invention de l'ingénieur moderne: L'École des Ponts et Chaussées, 1747–1851.* Paris: Presses de l'École national des Ponts et Chaussées.

Popplow, Marcus. 1998. *Neu, nützlich und erfindungsreich: Die Idealisierung von Technik in der frühen Neuzeit.* Münster: Waxmann.

———. 2006. "Unsichere Karrieren: Ingenieure und Techniker in Mittelalter und Früher Neuzeit 500–1750." In *Geschichte des Ingenieurs: Ein Beruf in sechs Jahrtausenden*, edited by Walter Kaiser and Wolfgang König, 71–125. Munich and Vienna: Hanser.

———, ed. 2010a. *Landschaften agrarisch-ökonomischen Wissens, Strategien innovativer Ressourcennutzung in Zeitschriften und Sozietäten des 18. Jahrhunderts.* Münster: Waxmann.

———. 2010b. "Die ökonomische Aufklärung als Innovationskultur des 18. Jahrhunderts zur optimierten Nutzung natürlicher Ressourcen." In *Landschaften agrarisch-ökonomischen Wissens, Strategien innovativer Ressourcennutzung in Zeitschriften und Sozietäten des 18. Jahrhunderts*, edited by Marcus Popplow, 2–48. Münster: Waxmann.

———. 2015. "Formalization and Interaction: Toward a Comprehensive History of Technology-Realted Knowledge in Early Modern Europe." *Isis* 106 (4): 848–856.

Poten, Bernhard von. 1896. *Geschichte des Militär- Erziehungs- und Bildungswesens in den Landen deutscher Zunge.* Vol. 4, *Preussen*. Berlin: A. Hofmann & Comp.

Preußische Akademie der Wissenschaften, ed. 1892–1982. *Acta Borussica, Denkmäler der Preußischen Staatsverwaltung im 18. Jahrhundert.* 42 vols. Berlin: Paul Parey.

Pugliano, Valentina. 2012. "Specimen Lists: Artisanal Writing or Natural Historical Paperwork?" *Isis* 103: 716–726.

Radder, Hans. 2009. "Science, Technology and the Science-Technology Relationship." In *Philosophy of Technology and Engineering Sciences*, edited by Anthonie Meijers, 65–91. Amsterdam: Elsevier.

Radkau, Joachim. 1989. *Technik in Deutschland: Vom 18. Jahrhundert bis zur Gegenwart.* Frankfurt a.M.: Suhrkamp.

Radkau, Joachim, and Ingrid Schäfer. 1987. *Holz: Ein Naturstoff in der Technikgeschichte.* Reinbek: Rohwohlt.

Radtke, Wolfgang. 1981. *Die Preußische Seehandlung zwischen Staat und Wirtschaft in der Frühphase der Industrialisierung.* Berlin: Colloquium Verlag.

Reich, Björn, Frank Rexroth, and Matthias Roick, eds. 2012. *Wissen maßgeschneidert: Experten und Expertenkulturen im Europa der Vormoderne.* Munich: Oldenburg Verlag.

Renn, Jürgen. 2020. *The Evolution of Knowledge: Rethinking Science in the Anthropocence.* Princeton: Princeton University Press.

Rexroth, Frank. 2012. "Systemvertrauen und Expertenskepsis: Die Utopie vom maßgeschneiderten Wissen in den Kulturen des 12. bis 16. Jahrhunderts." In *Wissen maßgeschneidert: Experten und Expertenkulturen im Europa der Vormoderne*, edited by Björn Reich, Frank Rexroth, and Matthias Roick, 12–44. Munich: Oldenburg Verlag.

Rheinberger, Hans-Jörg. 1997. *Towards a History of Epistemic Things: Synthesizing Proteins in the Test Tube.* Stanford: Stanford University Press.

Rieck, W. 1928. "Zur ältesten Geschichte der Tierärztlichen Hochschule Berlin." *Veterinärhistorisches Jahrbuch* 4: 118–119, 135–136.

Roberts, Lissa, Simon Schaffer, and Peter Dear, eds. 2007. *The Mindful Hand: Inquiry and Invention from the Late Renaissance to Early Industrialization.* Amsterdam: Koninklijke Nederlandse Akademie van Wetenschappen.

Roberts, Lissa, and Simon Werrett, eds. 2017. *Compound Histories: Materials, Governance, and Production, 1760–1840.* Leiden: Brill.

Rocke, Alan. 2018. "Theory versus Practice in German Chemistry: Erlenmeyer beyond the Flask." *Isis* 109 (2): 254–275.

Ropohl, Günter. 1997. "Allgemeine Technologie als Grundlage für ein umfassendes Technikverständnis." In *Allgemeine Technologie zwischen Aufklärung und Metatheorie: Johann Beckmann und die Folgen*, edited by Gerhard Banse, 111–121. Berlin: Edition Sigma.

———. 2006. "Allgemeine Technikwissenschaft." In *Erkennen und Gestalten: eine Theorie der Technikwissenschaften*, edited by Gerhard Banse, Arnim Grunwald, Wolfgang König, and Günter Ropohl, 331–341. Berlin: Edition Sigma.

Rostow, Walt Whitman. 1960. *The Stages of Economic Growth: A Non-Communist Manifesto*. Cambridge: Cambridge University Press.

Rudwick, Martin. 2005. *Bursting the Limits of Time: The Reconstruction of Geohistory in the Age of Revolution*. Chicago: University of Chicago Press.

Schaffer, Simon. 2007. "'The Charter'd Thames': Naval Architecture and Experimental Spaces in Georgian Britain." In *The Mindful Hand: Inquiry and Invention from the Late Renaissance to Early Industrialisation*, edited by Lissa Roberts, Simon Schaffer, and Peter Dear, 279–305. Amsterdam: Koninklijke Nederlandse Akademie van Wetenschappen.

Scharfenort, Louis von. 1910. *Die Königlich Preußische Kriegsakademie, 1810–1910, im dienstlichen Auftrag aus amtlichen Quellen dargestellt*. Berlin: Ernst Siegfried Mittler und Sohn.

Schatzberg, Eric. 2018. *Technology: Critical History of a Concept*. Chicago: University of Chicago Press.

Schauz, Désirée, and David Kaldeway. 2018. "Why Do Concepts Matter in Science Policy?" In *Basic and Applied Research: The Language of Science Policy in the Twentieth Century*, edited by David Kaldeway and Désirée Schauz, 1–32. New York: Berghahn Books.

Schellhas, Walter. 1959. "Alexander von Humboldt und Freiberg in Sachsen." In *Alexander von Humboldt 14. 9. 1769–6.5. 1859: Gedenkschrift zur 100. Wiederkehr seines Todestages*, edited by Alexander von Humboldt-Kommission der Deutschen Akademie der Wissenschaften zu Berlin, 339–422. Berlin: Akademie Verlag.

———. 1985. "Vom erzgebirgischen Bergjungen zum französischen Inspecteur général honoraire und Ritter der Ehrenlegion—Leben und Wirken von Johann Gottfried Schreiber (1746 bis 1827) in Sachsen und Thüringen." In *Leben und Wirken Deutscher Geologen im 18. und 19. Jahrhundert*, edited by Hans Prescher, 357–372. Leipzig: Deutscher Verlag für Grundstoffindustrie.

Schemmel, Matthias. 2008. *The English Galileo: Thomas Harriot's Work on Motion as an Example of Preclassical Mechanics*. Dordrecht: Springer.

Schickert, Otto. 1895. *Die Militärärztlichen Bildungsanstalten von ihrer Gründung bis zur Gegenwart*. Reprint, Zurich: Olms, 1986.

Schleiff, Hartmut. 2013. "Aufstieg und Ausbildung im sächsischen Bergstaat zwischen 1765 und 1868." In *Staat, Bergbau und Bergakademie: Montanexperten im 18. und frühen 19. Jahrhundert*, edited by Hartmut Schleiff and Peter Konečný, 125–159. Stuttgart: Steiner.

Schleiff, Hartmut, and Peter Konečný, eds. 2013. *Staat, Bergbau und Bergakademie: Montanexperten im 18. und frühen 19. Jahrhundert*. Stuttgart: Steiner.

Schröter, Alfred. 1962. "Die Entstehung der deutschen Maschinenbauindustrie in der ersten Hälfte des 19. Jahrhunderts." In *Die deutsche Maschinenbauindustrie in der industriellen Revolution*, edited by Alfred Schröter and Walter Becker, 11–133. Berlin: Akademie Verlag.

Schuster, John. 2017. "Consuming and Appropriating Practical Mathematics and the Mixed Mathematical Fields, or Being 'Influenced by Them': The Case of the Young Descartes." In *Mathematical Practitioners and the Transformation of Natural Knowledge in Early Modern Europe*, edited by Lesley B. Cormack, Steven A. Walton, and John A. Schuster, 37–68. Cham: Springer.

Sennewald, Rainer. 1994. "Das Lehrsystem in Freiberg, die Bildungsvorstellungen von Fr. A. von Heinitz in Preußen und Alexander von Humboldt." In *Beiträge zur Alexander-von-Humboldt-Forschung*, vol. 18, Studia Fribergensia, edited by Kurt-R. Biermann, Conrad Grau, and Christian Suckow, 289–301. Berlin: Akademie Verlag.

———. 2002. "Die Stipendiatenausbildung von 1702 bis zur Gründung der Bergakademie Freiberg 1765/66." In *Technische Universität Bergakademie Freiberg, Festgabe zum 300. Jahrestag der Gründung der Stipendienkasse für die akademische Ausbildung im Berg- und Hüttenfach zu Freiberg in Sachsen*, edited by Rektor der Technischen Universität Bergakademie Freiberg, 407–429. Freiberg: TU Bergakademie Freiberg.

Shapin, Steven. 1994. *A Social History of Truth: Civility and Science in Seventeenth-Century England*. Chicago: University of Chicago Press.

Siebeneicker, Arnulf. 2002. *Offizianten und Ouvriers: Sozialgeschichte der Königlichen Porzellan-Manufaktur und der Königlichen Gesundheitsgeschirr-Manufaktur in Berlin 1763–1880*. Berlin: de Gruyter.

Simon, Oskar. 1902. *Die Fachausbildung des Preussischen Gewerbe- und Handelsstandes im 18. und 19. Jahrhundert nach den Bestimmungen des Gewerbesrechts und der Verfassung des gewerblichen Unterrichtswesens*. Berlin: Heines.

Small, Alboin. 1909. *The Cameralists: The Pioneers of German Social Polity*. Chicago: University of Chicago Press.

Smith, Crosbie, and Norton M. Wise. 1989. *Energy and Empire: A Biographical Study of Lord Kelvin*. Cambridge: Cambridge University Press.

Smith, John Graham. 1979. *The Origins and Early Development of the Heavy Chemical Industry in France*. Oxford: Clarendon Press.

Smith, Pamela H. 2004. *The Body of the Artisan: Art and Experience in the Scientific Revolution*. Chicago: University of Chicago Press.

Smith, Woodruff D. 1991. *Politics and the Sciences of Culture in Germany 1840–1920*. New Yrok: Oxford Unversity Press.

Sokoll, Thomas. 2007. "Kameralismus." In *Enzyklopädie der Neuzeit*, edited by Friedrich Jaeger, vol. 6, 290–299. Stuttgart and Weimar: Metzler.

Spary, Emma C. 2010. "Liqueurs and the Luxury Marketplace in Eighteenth-Century Paris." In *Materials and Expertise in Early Modern Europe: Between Market and Laboratory*, edited by Ursula Klein and Emma C. Spary, 225–255. Chicago: University of Chicago Press.

———. 2013. *Eating the Enlightenment: Food and the Sciences in Paris, 1670–1760*. Chicago: University of Chicago Press.

Staudenmaier, John. 1985. *Technology's Storytellers*. Cambridge, MA: MIT Press.

Stewart, Larry. 1992. *The Rise of Public Science: Rhetoric, Technology, and Natural Philosophy in Newtonian Britain, 1660–1750*. Cambridge: Cambridge University Press.

Stichweh, Rudolf. 1992. *Zur Entstehung der wissenschaftlichen Disziplinen: Physik in Deutschland, 1740–1890*. Frankfurt a.M.: Suhrkamp.

Stoeller, Dr. 1800. "Nekrolog Johann Christian Wiegleb." *Allgemeines Journal der Chemie* 4: 684–720.

Straube, Hans Joachim. 1930. "Chr. P. Wilhelm Beuth." *Deutsches Museum, Abhandlungen und Berichte* 1930: 117–152.

Strecke, Reinhart. 2000. *Anfänge und Innovation der preußischen Bauverwaltung: Von David Gilly zu Karl Friedrich Schinkel*. Köln: Böhlau.

Strecke, Reinhart, Christof Baier, Elke Blauert, et al., eds. 2000. *Mathematisches Calcul und Sinn für Ästhetik: Die preußische Bauverwaltung 1770–1848, Ausstellung des Geheimen Staatsarchivs Preußischer Kulturbesitz in Zusammenarbeit mit der Kunstbibliothek der Staatlichen Museen zu Berlin Preußischer Kulturbesitz*. Berlin: Duncker & Humblot.

Stürzbecher, Manfred. 1966. *Beiträge zur Berliner Medizingeschichte: Quellen und Studien zur Geschichte des Gesundheitswesens vom 17. bis zum 19. Jahrhundert*. Berlin: de Gruyter.

Szabadváry, Ferenc. 1966. *Geschichte der Analytischen Chemie*. Braunschweig: Vieweg.

Teich, Mikuláš. 1975. "Born's Amalgamation Process and the International Metallurgical Gathering at Skleno in 1786." *Annals of Science* 32: 305–340.

Tenorth, Heinz-Elmar. 2012. "Geschichte der Universität zu Berlin, 1810–2010, Zur Einleitung." In *Geschichte der Universität Unter den Linden*, edited by Heinz-Elmar Tenorth, vol. 1, xv–xliii. Berlin: Akademie Verlag.

Tetzlaff, Claudia, and Hartmut Krohm. 2007. *KPM Welt, Ein Handbuch zur Ausstellung KPM-Welt*. Berlin: Königliche Porzellan-Manufaktur Berlin GmbH.

Timm, Albrecht. 1960. *Die Universität Halle-Wittenberg*. Frankfurt am Main: W. Weidlich.

———. 1964. *Kleine Geschichte der Technologie*. Stuttgart: W. Kohlhammer.

Tomic, Sacha. 2010. *Aux origines de la chimie organique: Méthodes et pratiques des pharmaciens et des chimistes, 1785–1835*. Rennes: Presses Universitaires de Rennes.

Tomory, Leslie. 2012. *Progressive Enlightenment: The Origins of the Gaslight Industry, 1780–1820*. Cambridge, MA: MIT Press.

———. 2014. "Science and the Arts in William Henry's Research into Inflammable Air During the Early Nineteenth Century." *Annals of Science* 71 (1): 61–81.

Trebra, Friedrich Wilhelm Heinrich von. 1785. *Erfahrungen vom Innern der Gebirge*. Dessau and Leipzig: Verlagskasse für Gelehrte und Künstler.

———. 1818. *Bergmeister-Leben und Wirken in Marienberg, vom 1. Decbr. 1767 bis August 1779*. Freyberg: Craz und Gerlach.

Tress, Werner. 2010. "Wissenschaft zwischen neuhumanistischem Bildungsideal und Staatsnutzen: Zur Gründung der Berliner Universität 1810." *Zeitschrift für Religions- und Geistegeschichte* 62: 261–281.

Treue, Wilhelm. 1984. *Wirtschafts- und Technikgeschichte Preussens*. Berlin: de Gruyter.

Troitzsch, Ulrich. 1966. *Ansätze technologischen Denkens bei den Kameralisten des 17. und 18. Jahrhunderts*. Berlin: Duncker & Humblot.

Trommsdorff, Johann Bartholomäus. 1793a. "Der vollkommene Apotheker." *Almanach oder Taschenbuch für Scheidekünstler und Apotheker* 1793: 75–96.

———. 1793b. "Plan und Zweck dieser Zeitschrift." *Journal der Pharmacie für Aerzte, Apotheker und Chemisten* 1, pt. 1: i–xii.

———. 1793c. "Methode, junge Leute zu brauchbaren Apothekern zu erziehen." *Journal der Pharmacie für Aerzte, Apotheker und Chemisten* 1, pt. 1: 29–39.

Turner, Steven R. 1971. "The Growth of Professional Research in Prussia, 1818 to 1848: Causes and Context." *Historical Studies in the Physical Sciences* 3: 137–182.

Valleriani, Matteo. 2010. *Galileo Engineer*. Dordrecht: Springer.

———, ed. 2017. *The Structures of Practical Knowledge*. Cham: Springer.

Vérin, Hélène. 1993. *La gloire des ingénieurs: l'intelligence technique du XVIe au XVIIIe siècle*. Paris: Albin Michel.

Vogel, Jakob. 2008. *Ein schillerndes Kristall: Eine Wissensgeschichte des Salzes zwischen Früher Neuzeit und Moderne*. Köln: Böhlau.

———. 2013. "Aufklärung untertage: Wissenswelten des europäischen Bergbaus im ausgehenden 18. und frühen 19. Jahrhundert." In *Staat, Bergbau und Bergakademie: Montanexperten im 18. und frühen 19. Jahrhundert*, edited by Hartmut Schleiff and Peter Konečný, 13–31. Stuttgart: Steiner.

Vogler, Günther, and Klaus Vetter. 1973. *Preußen: Von den Anfängen bis zur Reichsgründung.* Berlin: VEB Verlag der Wissenschaften.

Voigt, C. F. 1784. "Auszug eines Schreibens von Herrn Apotheker Voigt zu Erfurt." *Almanach oder Taschenbuch für Scheidekünstler und Apotheker* 1784: 184–190.

Vozár, Joseph, ed. 1983. *Das Goldene Bergbuch* (1764). Bratislava: Slovenskej Akadémie Vied.

Wagenbreth, Otfried. 1994. *Die Technische Universität Bergakademie Freiberg und ihre Geschichte dargestellt in Tabellen und Bildern.* Leipzig and Stuttgart: Deutscher Verlag für Grundstoffindustrie.

———. 1996. "Grubenrisse und geologische Karten als Hilfsmittel der Montanarchäologie." *Berichte der Geologischen Bundesanstalt* 35: 367–369.

Wakefield, Andre. 2009. *The Disordered Police State: German Cameralism as Science and Practice.* Chicago: University of Chicago Press.

Walcha, Otto. 1975. *Meißner Porzellan.* Gütersloh: Bertelsmann Lexikon Verlag.

Weber, Jutta. 2011. "Technoscience as Popular Culture: On Pleasure, Consumer Technologies, and the Economy of Attention." In *Science Transformed? Debating Claims of an Epochal Break,* edited by Alfred Nordmann, Hans Radder, and Gregor Schiemann, 159–176, Pittsburgh: University of Pittsburgh Press.

Weber, Max Maria Freiherr von. 1854. "Ueber Bildung der Techniker und deren Prüfung für den öffentlichen Dienst." *Der Civilingenieur* 1: 99–109.

Weber, Wolfhard. 1976. *Innovationen im frühindustriellen Bergbau und Hüttenwesen, Friedrich Anton von Heynitz.* Göttingen: Vandenhoeck & Ruprecht.

———. 2015. "Erschließen, gewinnen, fördern: Bergbautechnik und Montanwissenschaften von den Anfängen bis zur Gründung Technischer Universitäten in Deutschland." In *Geschichte des deutschen Bergbaus,* vol. 2: *Salze, Erze und Kohlen: der Aufbruch in die Moderne im 18. und frühen 19. Jahrhundert,* edited by Wolfhard Weber, 217–408. Münster: Aschendorff.

Wefeld, Hans Joachim. 2000. "Preußens erste Bauschule." In *1799–1999: Von der Bauakademie zur Technischen Universität Berlin, Geschichte und Zukunft,* edited by Karl Schwarz, 64–74. Ulm: Ernst und Sohn.

Wengenroth, Ulrich. 2006. "Intuitiv-heuristische Methoden." In *Erkennen und Gestalten: eine Theorie der Technikwissenschaften,* edited by Gerhard Banse, Arnim Grunwald, Wolfgang König, and Günter Ropohl, 133–144. Berlin: Edition Sigma.

Werner, Abraham Gottlob. 1787. *Kurze Klassifikation und Beschreibung der verschiedenen Gebürgsarten.* Dresden: Waltherische Hofbuchhandlung.

————. 1791. *Neue Theorie von der Entstehung der Gänge, mit Anwendung auf den Bergbau, besonders den freibergischen*. Freiberg: Gerlachische Buchdruckrei.

————. 1809. *New Theory on the Formation of Veins: with Its Application to the Art of Working Mines*. Translated by Charles Anderson. Edingburgh: Constable.

————. 1971. *Short Classification and Description of the Various Rocks*. Translated and with an introduction and notes by Alexander M. Ospovat, 97–144. New York: Hafner.

Werrett, Simon. 2010. *Fireworks: Pyrotechnic Arts and Sciences in European History*. Chicago: University of Chicago Press.

Wise, Norton M. 2018. *Aesthetics, Industry, and Science: Hermann von Helmholtz and the Berlin Physical Society*. Chicago: University of Chicago Press.

Wolff, Christian. 1734. *Vollständiges Mathematisches Lexicon, darinnen alle Kunst-Wörter und Sachen, welche in der erwegenden und ausübenden Mathesi vorzukommen pflegen deutlich erkläret*. Leipzig: Friedrich Gleditschens sel. Sohn.

Wulf, Andrea. 2015. *The Invention of Nature: Alexander von Humboldt's New World*. New York: Alfred A. Knopf.

Wutke, Konrad. 1913. *Aus der Vergangenheit des Schlesischen Berg- und Hüttenlebens: Ein Beitrag zur Preußischen Verwaltungs- und Wirtschaftsgeschichte des 18./19. Jahrhunderts*. Breslau: R. Nischkowsky.

Zilsel, Edgar. 2000. *The Social Origins of Modern Science*. With a foreword by Joseph Needham and introduction by Diederick Raven and Wolfgang Krohn, edited by Diederick Raven, Wolfgang Krohn, and Robert S. Cohen. Dordrecht: Kluwer.

Zimmermann, Carl Friedrich. 1744. "Vorrede." In *Kleine mineralogische und Chymische Schriften*, by Johann Friedrich Henkel. Vienna and Leipzig: J. F. Jahn. 2d ed. 1769.

————. 1745. "Der Nutzen und die Nutzung des Bergwerks, wie solche nach den politisch-ökonomischen Grundsätzen eines landesfürstlichen Kammerkollegium können betrachtet und verbessert werden." *Leipziger Sammlungen* 1: 408–448.

————. 1746. *Ober-Sächsische Berg-Academie, in welcher die Bergwercks-Wissenschaften nach ihren Grund-Wahrheiten untersuchet, und nach ihrem Zusammenhange entworffen werden*. 3 pts. Dresden and Leipzig: Hekel.

INDEX

Persons

Accum, Friedrich Christian, 205, 207, 219–221
Achard, Franz Carl, 9, 21–23, 25–29, 32, 33, 35, 64, 69, 102, 121, 129, 131, 176, 178, 180, 181, 209
Agricola, Georg, 85, 87, 100–101, 112, 151, 153, 170
Alberti, Leon Battista, 176
Alder, Ken, 4
Altenstein, Karl Freiherr vom Stein zum, 204

Bacon, Francis, 1, 170–172
Bartels, Christoph, 97, 162
Baumé, Antoine, 64
Becherer, Friedrich Christian, 198, 200, 201
Beckmann, Johann, 131, 172, 194, 211
Beguelin, Nicolas de, 22
Bélidor, Bernard Forest de, 180
Bergling, Friedrich, 35, 41, 63, 65–75, 77
Bergman, Torbern, 37, 39, 40, 248n14
Berthollet, Claude-Louis, 72, 166, 180–181, 267n23
Beuth, Christian Peter Wilhelm, 16, 205–207, 211–218, 221
Blair, Ann, 3
Born, Ignaz von, 122, 138, 167
Borsig, August, 212

Böttger, Johann Friedrich, 57
Boyle, Robert, 181
Brandt, Georg, 118–119
Brix, Adolf Ferdinand Wenceslaus, 207
Buch, Leopold von, 90, 255n10
Bückling, Karl Friedrich, 129
Buffon, Georges-Louis Leclerc, comte de, 112
Bülow, Ludwig Friedrich Victor Hans Graf von, 214, 270n33

Calvör, Henning, 162, 230
Carlowitz, Hannß Carl von, 23
Carny, Jean-Antoine, 180
Carrel, Alexis, 237
Castillion, Friedrich Adolf Maximilian Gustav, 126
Castillon, Jean, 22, 111
Chalmers, Alan, 176
Channell, David F., 234, 239–240
Chaptal, Jean-Antoine, 180
Charpentier, Johann Friedrich Wilhelm von, 93–94, 138, 154
Christian Friedrich Carl Alexander, margrave of Ansbach-Bayreuth, 133
Claiß, Johann Sebastian, 137
Cook, Harold, 79
Crell, Lorenz, 48–49, 53–55, 106
Crelle, August Leopold, 212
Cronstedt, Axel Frederic, 36–37

Topics

Transformations: Studies in the History of Science and Technology

Jed Z. Buchwald, general editor